INTERNATIONAL SERIES OF MONOGRAPHS IN
PURE AND APPLIED BIOLOGY

Division: **ZOOLOGY**

General Editor: G. A. Kerkut

Volume 31

THE AETIOLOGY OF
COMPRESSED AIR INTOXICATION
AND
INERT GAS NARCOSIS

OTHER TITLES IN THE ZOOLOGY DIVISION

General Editor: G. A. KERKUT

Vol. 1. RAVEN — *An Outline of Developmental Physiology*
Vol. 2. RAVEN — *Morphogenesis: The Analysis of Molluscan Development*
Vol. 3. SAVORY — *Instinctive Living*
Vol. 4. KERKUT — *Implications of Evolution*
Vol. 5. TARTAR — *The Biology of Stentor*
Vol. 6. JENKINS — *Animal Hormones—A Comparative Survey*
Vol. 7. CORLISS — *The Ciliated Protozoa*
Vol. 8. GEORGE — *The Brain as a Computer*
Vol. 9. ARTHUR — *Ticks and Disease*
Vol. 10. RAVEN — *Oogenesis*
Vol. 11. MANN — *Leeches (Hirudinea)*
Vol. 12. SLEIGH — *The Biology of Cilia and Flagella*
Vol. 13. PITELKA — *Electron-Microscopic Structure of Protozoa*
Vol. 14. FINGERMAN — *The Control of Chromatophores*
Vol. 15. LAVERACK — *The Physiology of Earthworms*
Vol. 16. HADZI — *The Evolution of the Metazoa*
Vol. 17. CLEMENTS — *The Physiology of Mosquitoes*
Vol. 18. RAYMONT — *Plankton and Productivity in the Oceans*
Vol. 19. POTTS and PARRY — *Osmotic and Ionic Regulation in Animals*
Vol. 20. GLASGOW — *The Distribution and Abundance of Tsetse*
Vol. 21. PANTELOURIS — *The Common Liver Fluke*
Vol. 22. VANDEL — *Biospeleology—The Biology of Cavernicolous Animals*
Vol. 23. MUNDAY — *Studies in Comparative Biochemistry*
Vol. 24. ROBINSON — *Genetics of the Norway Rat*
Vol. 25. NEEDHAM — *The Uniqueness of Biological Materials*
Vol. 26. BACCI — *Sex Determination*
Vol. 27. JØRGENSEN — *Biology of Suspension Feeding*
Vol. 28. GABE — *Neorosecretion*
Vol. 29. APTER — *Cybernetics and Development*
Vol. 30. SHAROV — *Basic Anthropodan Stock*

OTHER DIVISIONS IN THE SERIES ON
PURE AND APPLIED BIOLOGY

BIOCHEMISTRY

BOTANY

MODERN TRENDS
IN PHYSIOLOGICAL SCIENCES

PLANT PHYSIOLOGY

The Aetiology of
Compressed Air Intoxication
and
Inert Gas Narcosis

BY

P. B. BENNETT

*Royal Naval Physiological Laboratory,
Alverstoke, Hants, England*

PERGAMON PRESS
OXFORD · LONDON · EDINBURGH · NEW YORK
TORONTO · PARIS · BRAUNSCHWEIG

Pergamon Press Ltd., Headington Hill Hall, Oxford
4 & 5 Fitzroy Square, London W.1

Pergamon Press (Scotland) Ltd., 2 & 3 Teviot Place, Edinburgh 1

Pergamon Press Inc., 44-01 21st Street, Long Island City, New York 11101

Pergamon of Canada, Ltd., 6 Adelaide Street East, Toronto, Ontario

Pergamon Press S.A.R.L., 24 rue des Écoles, Paris 5ᵉ

Friedr. Vieweg & Sohn Verlag, Postfach 185, 33 Braunschweig, West Germany

Copyright © 1966
Pergamon Press Ltd.

First edition 1966

Library of Congress Catalog Card No. 65-22882

PRINTED IN GREAT BRITAIN BY PAGE BROS. (NORWICH) LTD., NORWICH

(2416/66)

CONTENTS

	Acknowledgements	ix
	Preface	xi
	Introduction	xiii
I.	Signs and Symptoms	1
II.	The Causes of the Narcosis	19
III.	Tissue Carbon Dioxide Retention as the Cause of Inert Gas Narcosis	31
IV.	Mechanisms of the Narcosis	42
V.	The Electrical Activity of the Brain and Inert Gas Narcosis	65
VI.	The Possible Action of Inert Gases on Synaptic Mechanisms	81
VII.	Prevention of the Narcosis	86
VIII.	Addendum—Recent Developments	92
	References	100
	Author Index	107
	Subject Index	111

TO MY WIFE

ACKNOWLEDGEMENTS

GRATEFUL appreciation is extended to the following for permission to reproduce material.

Acta Physiologica Scandinavica; American Physiological Society; American Psychological Assoc.; Bureau of Medicine and Surgery U.S.N.; Elsevier Publishing Co.; Exerpta Medica Foundation; J. B. Lippincott Co.; Macmillan and Co. Ltd.; Navy Department, Ministry of Defence; Pergamon Press; The Biochemical Journal; The Rockefeller Institute Press; The Williams and Wilkins Co.; The Wistar Institute of Anatomy and Biology; U.S. Naval Medical Research Labs.; U.S. Naval Missile Center; U.S. Navy Experimental Diving Unit; U.S. School of Aerospace Medicine; Dr. J. Adolfson, Dept. of Psychology, Gothenburg, Sweden; Dr. G. Albano, "Ottavio Zanca", Palermo, Sicily; Surgeon Lieutenant Commander E. E. P. Barnard, Royal Naval Physiological Laboratory, Alverstoke, England; Professor J. W. Bean, University of Michigan, U.S.A.; Dr. A. R. Behnke, University of California Medical Center, U.S.A.; Dr. F. G. Carpenter, Dartmouth Medical School, U.S.A.; Professor R. M. Featherstone, University of California, U.S.A.; Professor W. O. Fenn, University of Rochester, U.S.A.; Mr. H. V. Hempleman, Royal Naval Physiological Laboratory, Alverstoke, England; Dr. C. M. Hesser, Karolinska Institute, Sweden; Dr. Alma Howard, Christie Hospital and Holt Radium Institute, England; Dr. Jean Marshall, Harvard Medical School, U.S.A.; Commander E. M. Neptune, M.C., U.S. Naval Medical Research Unit, San Francisco, U.S.A.; Professor C. B. Pittinger, Vanderbilt University, U.S.A.; Dr. H. R. Schreiner, Union Carbide Co., Tonawanda, U.S.A.; Captain C. W. Shilling, M.C., U.S.N., The George Washington University, U.S.A.; Captain R. D. Workman, M.C., U.S. Naval Experimental Diving Unit, Washington, U.S.A.

PREFACE

IT IS now over 100 years since it was first noted that the air we breathe is capable of producing narcosis and anaesthesia, if it is breathed at sufficient increased pressure. With the increasing interest in the natural resources to be found in and beneath the sea and the advent of renewed interest in diving to great depths, there has been revived interest in this phenomenon over the past decade.

Further, the discovery that the noble gases will also produce similar signs and symptoms, although considered chemically inert, has aroused the hope that here may be the key to anaesthesia and the manner of its production.

The literature, as a consequence, has grown considerably in recent years. It is believed that this justifies an attempt to survey the present knowledge on the subject and to analyse and correlate the increasing information emanating from many countries.

INTRODUCTION

ONE of the most baffling problems in medicine is the manner by which anaesthetics induce narcosis and anaesthesia especially in relation to that caused by volatile anaesthetics such as ether, nitrous oxide and cyclopropane. These seemingly inert substances are in some way able to induce profound unconsciousness without apparently entering into any direct biochemical process.

Perhaps even more puzzling is the fact that the very air we breathe is capable of producing exactly the same conditions if the partial pressure is sufficiently high and further that all the so-called "noble" gases of the helium group elements of the Periodic Table are potent narcotics inducing surgical anaesthesia at an appropriate partial pressure.

It is the purpose of this monograph to consider the narcotic effects in particular produced by these so-called inert gases and air.

Anaesthesia and Narcosis

In the chapters that follow, the terms anaesthesia and narcosis will be appearing frequently. It is therefore pertinent to consider exactly what we mean by these terms.

The term narcosis is derived from the Greek word *narke*—numbness, which Galen used to describe a particular group of drugs including opium. The Oxford Dictionary definition of opium as "a substance, which, when swallowed, inhaled or injected into the system, induces drowsiness, sleep, stupefaction or insensibility according to its strength and amount taken" is as may be seen a definition of the action of opium. The word narcosis is therefore generally used for a state similar to that produced by opium alkaloids.

The term anaesthesia is only of comparatively recent origin. It was coined in the nineteenth century and is attributed to Oliver Wendell Holmes. He wrote a letter to Morton, the discoverer of ether, suggesting the words anaesthesia, anaesthetic and anaesthetist. These words were related to insensibility, in particular to touch (*anaisthetos* is Greek for "insensible").

Goodman and Gilman (1943) define anaesthesia as "the loss not only of all modalities of sensation but also of consciousness", whereas narcosis is defined as a "condition of analgesia accompanied by deep sleep or stupor. The action of a narcotic drug differs in that the pain is relieved before the occurrence of sleep or unconsciousness." Perhaps the most pertinent comment was made in 1950 by Thomas Butler, in his review of the theories of general anaesthesia, who suggests that "the vagueness and multiplicity of meanings are so firmly established in the usage of 'narcosis' and 'narcotic' in English, as to render futile any insistence on rigid definitions".

The first chapter of this monograph is therefore confined to describing the signs and symptoms caused by exposing man and animals to increased pressures of air and mixtures of oxygen and the so-called noble or rare gases. In doing so it is hoped that the reader will form some opinion as to the meaning of the term "narcosis" under the above conditions.

So far as the term "inert gas" is concerned in relation to the noble gases there is probably no better definition than that of Featherstone and Muehlbaecher (1963) that it is used "in a metabolic sense and includes those gases which are generally considered to exert their biological effects without undergoing any change in their own chemical structures or modifying the primary chemical structure of other substances".

Increased Pressures of Air and Inert Gases

As much of the work to be considered is at pressures in excess of 1 atmosphere it may be of value to first consider some basic information. It will be taken that at sea level man is exposed to an air pressure of 760 mm Hg or 14·69 lb/in^2. This is called the pressure of *1 atmosphere* or *atmospheric pressure*. One atmosphere is composed of 78·084 per cent nitrogen, 20·946 per cent oxygen, 0·934 per cent argon, 0·033 per cent carbon dioxide and a very small quantity, 0·003 per cent, of other members of the rare gas series.

There are two ways by which pressure is signified. One includes the pressure exerted by the atmosphere and is called the *absolute pressure*. The other does not take this into account and is called simply *pressure* or, as this is usually the pressure registered on pressure gauges, the *gauge pressure*.

At some 33 feet of sea water the pressure exerted due to the weight of water is equivalent to that produced by 1 atmosphere of air. We may therefore say that 33 feet is equivalent to a pressure of 1 atmosphere or approximately 15 lb/in^2. Alternatively 33 feet is equivalent to a pressure of 2 atmospheres absolute or 30 lb/in^2 absolute.

Now Boyle's law states that the volume of a gas varies inversely as the absolute pressure, while the density varies directly as the absolute pressure, i.e. Pressure $\times$ Volume $=$ Constant. The lungs may be likened to a balloon and in accordance with Boyle's law it is apparent that increasing the pressure surrounding the balloon or lungs to 2 atmospheres absolute must result in a corresponding reduction of the volume by one-half. At 100 feet or 4 atmospheres absolute the lung volume is reduced to one-quarter.

It is only very well-trained individuals, such as naked pearl and sponge divers, who are able to withstand such a severe reduction in lung volume without tissue damage. Obviously to swim naked, i.e. without breathing equipment, to depths in excess of 100 feet is impossible. At 200 feet or 7 atmospheres absolute the lung volume would be reduced to one-seventh of normal. It is possible for the lung volume to be reduced from its normal 6 litres to 2 litres without damage. Most normal individuals should therefore be able to attain a depth of about 60 feet without trouble. That pearl divers are able to dive deeper (90 feet) is due to their greater lung volumes, which are normally in the region of 8 litres.

In order to dive below 100 feet it is therefore necessary to supply air to the lungs at a pressure equivalent to that surrounding the lungs. This requires the use of diving suits. Under these conditions the lung volume is unaltered but the air in the lungs is at increased pressure. Due to this increased pressure there is a consequent increase in the respective *partial* pressures of the gases in the lungs. In accordance with Dalton's law of partial pressures the *partial pressure* of each gas in a mixture is equal to the pressure which it would exert if it alone occupied the total volume which that mixture occupies. It is the increased partial pressures of the gases in the air which is responsible for the many hazards found in diving.

This is because the process of evolution has ensured that man is adapted to life at approximately 1 atmosphere absolute. Any marked variation from this constitutes a hazardous environment to man. For example, exposure to high altitude will result in oxygen lack or hypoxia and perhaps decompression sickness commonly known as "the bends". The first condition is the result of the reduced partial pressure of oxygen as a result of the reduced total pressure, and the second either to the rate of ascent or the altitude itself being so great that the nitrogen equilibrated in the body at 1 atmosphere absolute is unable to leave the body with sufficient speed. Instead, like removing the top off a soda water bottle, bubbles are formed in the tissues. As a result and depending on their situation limb pains may occur, usually at the joints, or alternatively impairment of the function of the brain and spinal cord may result.

Decompression sickness is also found in divers and caisson workers (Fulton 1951). Due to the time spent breathing air at increased pressure the tissues slowly become saturated with gas at the increased pressure. If the pressure is released too quickly then, as with aviators, bubbles may occur in the tissues and "the bends" result.

The bends are often confused with depth intoxication or narcosis. They are however quite different. The bends only occur as a result of too rapid a decompression from exposure to increased pressures whereas depth intoxication is a narcotic state found in men and animals breathing air or inert gas–oxygen mixtures whilst actually *at* the increased pressure.

Depth intoxication should also not be confused with oxygen poisoning, which is the result of too high an oxygen partial pressure in the air (Bean 1945). Exposure to pure oxygen at pressures in excess of 2 atmospheres absolute results in epileptiform convulsions or, in the case of long exposure at even lower pressures, lung damage.

Metric Equivalents

Although it is customary in the U.S.A. and Great Britain to use the terms lb/in^2 or psi (pounds per square inch) or atm (atmospheres) many countries use the metric system. It may therefore assist the reader to know the following equivalents:

$$1 \text{ atmosphere} = 1 \cdot 033 \text{ kg/cm}^2$$
$$3 \cdot 28 \text{ feet} = 1 \text{ metre}.$$

CHAPTER I

SIGNS AND SYMPTOMS

THIS chapter is devoted to a description of the signs and symptoms of narcosis in men and animals breathing air or mixtures of oxygen and the noble or rare gases. To obtain a clear conception of these is difficult. The most useful analogy for the reader is the similarity of the signs and symptoms with alcoholic intoxication.

Although many workers have endeavoured to describe the sensations due to the narcosis, the most profitable information has been obtained from efforts to quantify the narcosis by means of psychometric tests. These however have certain defects in that learning and variations in motivation can give rise to misleading results. The situation is often hindered by imprecise information as to the pressure and time at pressure at which the tests were used.

Various animals and insects have been subjected to pressure as an alternative to man but such creatures usually require exposure to very high pressures before they show some sign or symptom of narcosis. It is also often difficult to assess by visual appraisal alone whether or not an animal or insect is narcotic.

In spite of these reservations, the literature available on this aspect of inert gas narcosis does serve as an indication of the meaning of the term narcosis as it is used in this field. Much work however remains before a logical appraisal of the signs and symptoms of narcosis is possible from the diverse and sometimes imprecise literature at present available.

Compressed Air Narcosis in Man

In 1861 Green, an American, was the first to note that breathing compressed air whilst diving produced a "feeling of sleepiness". At 160 feet (5·8 atmospheres absolute) he also noted hallucinations and an impairment of judgement. Green suggested that the signs and symptoms were of sufficient severity to merit an immediate return to atmospheric pressure. Paul Bert (1878) in his classic book *La Pression Barométrique* also reports briefly on the narcotic condition of divers at great depths. Similar signs and symptoms were reported in caisson workers at 150 feet (5·5 atmospheres absolute) by Hill and McLeod (1903). Somewhat later, Damant (1930) reported that at 300 feet (10 atmospheres absolute) men become mentally abnormal and suffer memory defects.

At this time the Admiralty appointed a committee, which included Sir Leonard Hill and Sir Robert Davis, to investigate the practical and material problems associated with deep diving and submarine escape. After 3 years of study and experiments a 176 page report resulted (1933). Included in this report was a section devoted to "semi-loss of consciousness". This described various signs and symptoms of narcosis and lapses of consciousness in 17 of 58 dives to 200–350 feet (7–11·6 atmospheres absolute). A considerable amount of time was spent in investigating this phenomenon. The condition was regarded as serious, as the diver would continue to give all the normal hand signals at depth but after decompression could not remember any of the events which took place while he was at pressure. All of the divers regained full consciousness during the decompression.

Some improvement in the condition was obtained if the diving helmet was flushed with air, which suggested that carbon dioxide might be responsible. Lights were lowered to see if it was due to the divers working in darkness but again this did not eradicate the trouble. Much speculation resulted as to the cause. As the trials progressed it became evident that some men were more affected than others and it was concluded that the question was probably mainly a matter of the individual.

At similar pressures, Hill and Phillips (1932) reported a dangerous overconfidence, with a dulling of mental ability and difficulty in retaining information and the ability to make rapid decisions. The authors of this paper excluded carbon dioxide as a possible factor and suggested that the cause could be impurities in the air from the compressors or that the signs and symptoms were of psychological origin.

Behnke, Thomson and Motley, in 1935, were the first to attribute the narcosis to the raised partial pressure of the physiologically inert gas nitrogen in the air. The narcosis was characterized as "euphoria, retardment of the higher mental processes and impaired neuromuscular co-ordination". At increased pressures of 4 atmospheres, laughter and loquacity were apparent, although with increased effort at self-control it was possible to overcome these symptoms. The response to visual, auditory and olfactory stimuli was delayed and there was a tendency to idea fixation as often seen in hypoxia. Errors were made in arithmetic and recording data. For example, 43 minutes was confused with 48 minutes and 12:15 written as 15:15. The ability to carry out fine movements was affected due to over-exaggeration of movement. If such movements were however carried out slowly, this exaggeration could be overcome. At 10 atmospheres stupefaction resulted and it was considered that unconsciousness would occur between 10 and 15 atmospheres.

Cousteau (1953), in his book *The Silent World*, gives a graphic account of nitrogen narcosis, which, not knowing of the earlier work of Behnke, he called "l'ivress des grandes profondeurs" (Rapture of the Deep). The hazard of this state to deep divers is illustrated in the description of how a colleague, Maurice Fargues, lost his life. He attained a depth of 396 feet whilst breathing compressed air but due to the narcosis at this depth had probably become unconscious, lost his mouthpiece and drowned. Cousteau suggests that the wide variation in susceptibility between individuals is due to differences in emotional stability. The more an individual is emotionally stable, the better he is able to counteract the narcosis by increased effort, until, if the pressure is high enough, consciousness is lost.

The signs and symptoms of the narcosis are very similar to those seen in alcoholic intoxication and the early stages of hypoxia and anaesthesia. Many factors other than the depth or pressure influence the severity of the narcosis. Alcoholic excess, fatigue, hard work (Adolfson 1964) and apprehension facilitate the condition. Frequent exposure to high pressure produces acclimatization.

Effect of Compressed Air on Tests of Performance in Man

The first quantitative study to demonstrate the type and degree of impairment of the conscious state by compressed air was made in 1937 by Shilling and Willgrube. At increased pressures of air between 3·7 and 10 atmospheres absolute (90–300 feet), 46 men were required to answer simple arithmetical problems: one each of addition, multiplication, subtraction and division.

The number of errors and the time taken to complete the test were noted. Additional tests were number cancellation, where in a sheet of numbers the subject was required to cross out as many of a given number as possible in 1 minute, and measurement of the reaction time in response to a sudden light stimulus.

As may be seen in Table 1, quantitative evidence of the slowing of normal mental and neuromuscular responses was obtained. The quantitative evidence was also found to support earlier suggestions that experience in work at increased pressures decreased the severity of the narcosis. A correlation was observed between mental ability and performance failure at pressure. Individuals of high mental ability were found to be less sensitive to the narcosis than those of low mental ability. The greatest narcotic effect was noted immediately upon reaching pressure and rapid compression caused dizziness. As a result, Shilling and Willgrube considered that the high nitrogen partial pressure could not be responsible since the tissue tension would not be sufficient in such a short time. This is however considered more fully in a later chapter. It will be seen in Table 1 that at 90 feet the time to carry out an arithmetic test was longer than at 100 feet, whereas the reverse is the usual occurrence. This was due to this being the first experience at pressure for many of the subjects and illustrates the different sensitivities of novices and trained divers.

In 1941 Case and Haldane published the results of many experiments in human physiology at high pressure. Included in their studies were the effect of various nitrogen–oxygen partial pressures, carbon dioxide, cold and the breathing of argon–oxygen, helium–oxygen and hydrogen–oxygen mixtures at increased pressures.

Manual skill and intellectual ability were examined at increased pressures of 8·6 and 10 atmospheres of air (250 feet and 300 feet). As a test of manual skill, steel balls were required to be placed in three holes with various implements. Intellectual ability was tested with a series of four-figure multiplications, e.g. 9746 × 4956. The time taken and the number of problems correct were noted.

At 250 feet, of seven subjects, two were distressed and felt

TABLE 1. EFFECT OF PRESSURE ON PSYCHOMETRIC TESTS (SHILLING AND WILLGRUBE 1937)

Pressure (ft)	0	90	100	125	150	175	200	225	250	275	300
Mean additional time to solve problems (sec)	0.35	11.09	6.89	7.65	9.74	11.95	13.98	17.17	26.07	26.53	31.42
Mean additional errors in solving problems	0.18	0.86	0.49	0.42	0.72	0.84	1.22	0.88	2.18	2.66	3.02
Mean decrease in numbers crossed out	—	−0.59	−0.09	−2.26	−2.30	−2.49	−2.55	−4.24	−5.85	−6.43	−8.74
Average reaction time (sec)	0.214	—	—	—	0.237	—	0.242	—	0.248	—	0.257
Mean additional time to solve problems (acclimatized subjects)	1.64	2.55	3.42	3.91	4.66	8.00	11.75	15.73	16.33	17.09	24.36

faint, one was euphoric, talkative and overconfident, and one elated. The remainder showed no obvious emotional reaction, even after 24 minutes at pressure. The manual dexterity test at this pressure showed no significant deterioration but the arithmetic test did show evidence of mental impairment. The time taken to answer the problems was usually increased. In a total of 33 problems four of the subjects increased their errors substantially from 6 at atmospheric pressure to 22 at the increased pressure.

At the higher pressure of 10 atmospheres absolute (300 feet), the effects were more apparent. Practical activity and judgement were severely impaired and the authors concluded that "no great trust should be placed in human intelligence under these circumstances". The narcosis was apparent during compression and reached a maximum within 2 minutes. After 30 minutes at the same pressure the narcosis was apparently no worse, indeed if anything there was a slight acclimatization subjectively.

Decompression resulted in an abrupt recovery of faculties, usually at some 5 atmospheres pressure. Case and Haldane concluded that the narcosis principally affected intellectual and moral performance. Muscular skill was not unduly impaired. The results of *ad hoc* experiments, in which the effect of the partial pressure of nitrogen was examined at decreased oxygen partial pressures, led to the further conclusion that there is little if any synergism between high nitrogen partial pressures and oxygen lack.

Rashbass (1955) confronted subjects with a list of sums consisting of two-figure by one-figure multiplication omitting noughts, ones, fives and multiples of 11, the score being the number correct in 2 minutes. At 8·6 atmospheres absolute of air (250 feet), Rashbass found a significant deterioration in performance. The mean number correct for 26 subjects at atmospheric pressure was 24·12 compared with 16·81 at pressure. Bennett and Glass (1961) used the same test at 7 atmospheres absolute (200 feet) and observed a deterioration from 20·34 to 15·67 problems at pressure. A letter cancellation test at 200 feet was not significantly affected.

An attempt to use more complicated tests was made by Rosenberg and Ramsdell (1957). They used complex measuring instruments such as the "Porteus Maze", "Goldstein–Scleerer test" and digit substitution. Although they did demonstrate a performance decrement at 4 atmospheres absolute (100 feet), their conclusions were not free from learning factors associated with the instruments.

More conclusive evidence of impaired performance at relatively low nitrogen partial pressures was obtained by Kiessling and Maag (1962). Ten subjects were given a choice reaction time test, a complex mechanical dexterity test evolved from the Purdue Pegboard (Buros 1949) and a difficult conceptional reasoning test requiring the classification of 32 small wooden blocks. The tests were carried out at a pressure of 4 atmospheres absolute (100 feet) of air and at atmospheric pressure. Evidence was obtained that the neural systems responsible for reasoning and immediate memory are affected by inert gases to a greater extent than those supporting simple motor co-ordination and choice reactions. The authors pointed out that the degree of performance impairment is largely a function of the test used. This may appear self-evident but many workers ignore this fact and make dogmatic conclusions as to the depth or pressure at which narcosis first appears. Thus the critical pressure for the onset of narcosis varies with different workers between 100 and 200 feet.

If psychometric tests are used in screening men for sensitivity at increased pressures the tests must therefore be selected with reference to the specific tasks required of the men whilst at pressure. Although Kiessling and Maag have criticized the use of such tests as letter cancellation, tests of gross motor skill or well-practised tasks such as multiplication of two digits by one, on the grounds that they are merely a test of memory of the multiplication tables, the author believes the latter test (Rashbass 1955) one of the most sensitive and reliable available.

The tests used by Kiessling and Maag showed that at 100 feet the performance of reasoning ability was decreased by 33·46 per cent compared with the efficiency at atmospheric pressure. For reaction time the decrement was 20·85 per cent and for mechanical dexterity 7·9 per cent (Table 2). Duration of exposure did not appear to materially affect the extent of performance impairment. During the first 12 minutes at 100 feet, the mean combined test performance impairment was 19·3 per cent compared with 21·2 per cent for the period 12–24 minutes and 22·1 per cent for 24–36 minutes.

The experiments described previously were all carried out

during simulated dives in compression chambers. Miles and Mackay (1959) applied arithmetical and memory tests to divers under the water in diving suits at pressures of 4–6·5 atmospheres absolute (100–180 feet). No significant difference was found in the ability to complete the tests. This result, which is in disagreement

TABLE 2. TESTS ON PERFORMANCE AT SEA LEVEL AND 100 FEET
(KIESSLING AND MAAG 1962)

		Atm. Press.	100 ft	$S.E._{dm}$	t
Reaction time	Mean	23·74	28·69	0·728	6·8*
(1/100 sec)	S.D.	4·86	6·38		
Mechanical	Mean	28·09	25·87	0·500	4·44*
dexterity	S.D.	2·84	2·07		
(pieces assembled)					
Conceptual	Mean	7·68	10·25	0·468	5·491*
reasoning	S.D.	2·07	2·82		
(sec per problem)					

* Significant at greater than 0·01 level.

with most evidence obtained from compression chamber experiments, may be due to the nature of the tests used or that the subjects were well-trained divers. Single subjects did show marked evidence of impaired performance which was not apparent when the means for all the subjects were examined.

Effect of Various Nitrogen–Oxygen Partial Pressures on Psychomotor Performance

Albano, Criscuoli and Ciulla (1962) investigated the effect of breathing mixtures of different nitrogen–oxygen partial pressure on psychometric tests. Arithmetic and tests of visual memory and imagination were given to 7 subjects at 10 atmospheres absolute (300 feet). The two mixtures were air and 96 per cent nitrogen–4 per cent oxygen. Care was taken to ensure by preliminary training that the men had attained a stable level of efficiency and had reached their learning plateau. This is another most important consideration which many workers ignore in experiments of this kind. The arithmetic test consisted of 10 problems in simple arithmetic to be worked out to 5 figures. A standard time of $2\frac{1}{2}$ minutes was allowed. The score was based on the number of figures the subject was able to multiply and the percentage of

TABLE 3. ARITHMETIC TEST RESULTS AT 10 ATMOSPHERES ABSOLUTE (ALBANO et al. 1962)

	Figures multiplied			Percentage of errors			Difference	
Subject	(1) Ambient pressure	(2) 10 atm air	(3) 10 atm 96% N_2–4% O_2	(4) Ambient pressure	(5) 10 atm air	(6) 10 atm 96% N_2–4% O_2	(5)–(4)	(6)–(5)
A.G.	23	18	12	4·35	22·2	41·6	17·85	19·4
P.V.	24	19	15	4·25	79	86·6	74·75	7·6
R.S.	50	43	33	—	23	21·8	23·00	–1·2
M.E.	40	20	14	10	30	42·8	20·00	12·8
S.V.	36	32	28	28	53·6	71·4	25·60	17·8
C.B.	27	24	20	7·4	50	60	42·60	10·0
C.U.	45	34	30	—	26·4	30	26·40	3·6
							$M =$ 32·88	10·0
							$t =$ 4·30	4·20
							$P =$ <0·01	<0·01

mistakes. As may be seen in Table 3, there was a reduction in intellectual capacity at increased air pressures which was worse when the 96 per cent nitrogen–4 per cent oxygen mixture was breathed. In all the experiments there was a definite deterioration in handwriting at pressure, as has also been reported by Case and Haldane (1941) and Cabarrou (1964). It was concluded that the nitrogen partial pressure plays a fundamental role in the determination of narcotic changes in men at high pressures of air. Subjectively, all the subjects showed signs of euphoria, hyperexcitability and a reduction of intellectual–perceptive faculties

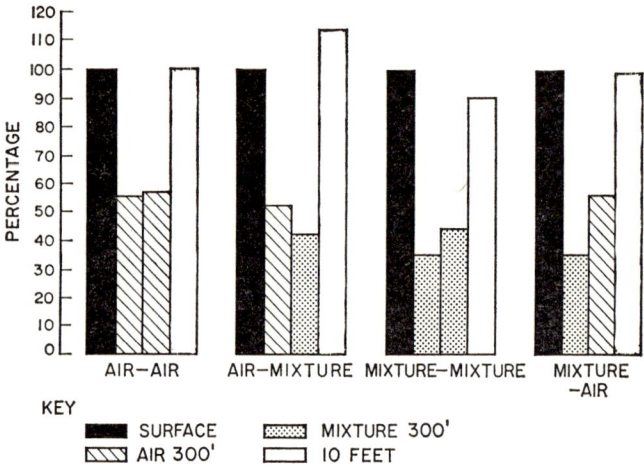

Fig. 1. Histogram of relative percentages of the number of simple arithmetic sums correct for 10 subjects at atmospheric pressure, 300 feet breathing air and 300 feet breathing a mixture of 95 per cent nitrogen–5 per cent oxygen (Barnard *et al.* 1962).

when breathing air. When breathing the mixture there was a feeling of well-being, physical relaxation and apathy. Albano *et al.* (1962) concluded that in the former condition the narcotic state produced by the nitrogen was overshadowed by the effects of the associated high oxygen partial pressure, with superimposition of the hyperexcitability and euphoric state.

Similar experiments were carried out by Barnard, Hempleman and Trotter (1962). They also found that there is an impairment in performance of individuals at simple arithmetic at 10 atmospheres absolute (300 feet) of air, which is worse if a mixture of

95 per cent nitrogen–5 per cent oxygen is breathed (Fig. 1). In agreement with Case and Haldane (1941) the narcosis was found to reach its maximum within 2 minutes of starting compression. Recovery was rapid during decompression. In confirmation that nitrogen is primarily responsible for the narcotic effect of air at high pressure, Barnard *et al.* (1962) also demonstrated that there was an inverse relationship between performance at arithmetic and the nitrogen partial pressure of the mixture breathed (Fig. 2).

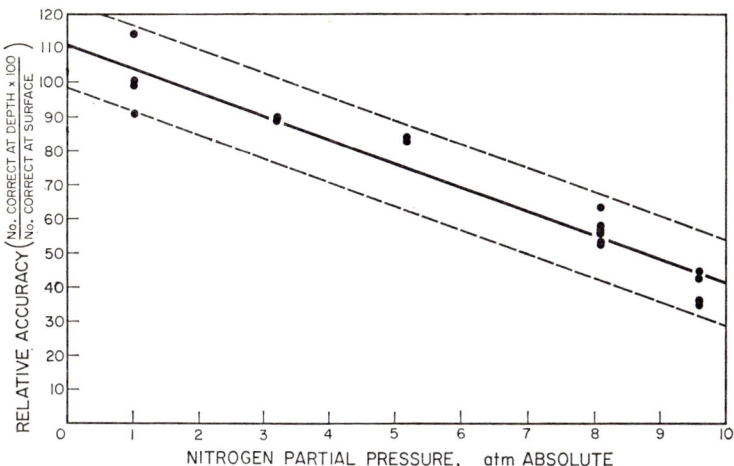

Fig. 2. Ability to answer arithmetical problems related to the nitrogen partial pressure to which the subjects were exposed (Barnard *et al.* 1962).

Some of the most valuable and most carefully controlled studies of the effect of increased pressures of air on human behaviour have been carried out recently by Hesser and his colleagues at the Institute of Psychology, University of Gothenburg (Frankenhaeuser, Graff-Lonnevig and Hesser 1960, 1963, Hesser 1963, Adolfson 1964). Like Albano *et al.* (1962) and Barnard *et al.* (1962), Frankenhaeuser *et al.* investigated the effect of different oxygen–nitrogen mixtures on the psychomotor function of 12 subjects but the nitrogen partial pressure was kept constant and only the oxygen varied. The tests were simple and four-choice reaction time and mirror drawing. In the latter test the subject was required to move a stylus along a path cut in a metal plate in the form of a five-pointed star which was visible in a mirror. Notches in the track tended to obstruct the stylus movement.

The score was the time to complete one run and the error score the time of contact between the stylus and the metal track. Five gas mixtures were used and the results are shown in Table 4. The degree of impairment at 5 atmospheres absolute (130 feet) is similar to that reported by Case and Haldane (1941) but considerably less than that found by Shilling and Willgrube (1937) and

TABLE 4. MEAN PERFORMANCE OF 12 SUBJECTS AT PSYCHOMOTOR TESTS DURING EXPOSURE TO A CONSTANT NITROGEN PARTIAL PRESSURE BUT VARYING OXYGEN PRESSURE (HESSER 1963, FRANKENHAEUSER et al. 1963)

Test		Partial pressures (atm abs.)				
	O_2	0·2	0·94	0·22	1·03	2·60
	N_2	0·74	—	3·92	3·91	3·94
Simple reaction (sec)		0·243	0·242	0·241	0·248	0·256
Choice reaction (sec)		0·671	0·683	0·685	0·691	0·698
Mirror drawing time (sec)		9·16	9·25	9·47	9·24	8·93
Mirror drawing error (sec)		2·89	2·85	3·39	3·11	3·34

Kiessling and Maag (1962). The reasons for the discrepancy are not clear. Hesser (1963) suggests that the reason could be the presence of small quantities of carbon dioxide in the pressure chamber in the latter two experiments or differences in individual susceptibility. Whatever the reason, the results do illustrate the difficulty of obtaining repeatable objective measurements of performance impairment due to increased pressures of air.

More important however is the fact that the degree of performance impairment increases with the increase in oxygen partial pressure when the nitrogen partial pressure is constant. This is more marked at high nitrogen partial pressures than at low (Frankenhaeuser et al. 1960). Hesser (1963) concluded that this effect was due to a corresponding increase in tissue carbon dioxide and that the carbon dioxide has a synergistic narcotic action. These points are discussed in detail in relation to other experiments in later chapters, particularly in Chapters III and IV.

It is clear however from the studies of Hesser et al. that the relationship between the increased nitrogen partial pressure, the increased oxygen partial pressure and the presence of carbon dioxide is very important in determining the extent of narcosis and consequent impairment of performance found for a given pressure. Considerable care is therefore required in ensuring that in studies of the narcosis only one of these variables is examined at any given time, the rest being kept constant. The effect of the increased gas density with its effect of increasing breathing resistance is a further factor and is also considered in detail in the chapters that follow. The increased density also causes carbon dioxide retention and adds further to the narcosis, especially if hard work is involved.

An excellent study by Adolfson (1964) of human behaviour at 4, 7, 10 and 13 atmospheres clearly indicates that hard work does indeed markedly increase the decrement in performance of an arithmetical task (Table 5) and manual dexterity. Adolfson also

Table 5. Mean percentage impairment in ability of 14 subjects to do an arithmetical test during rest and work on a bicycle ergometer (300 kg/min) (Adolfson 1964)

Absolute air pressures	4 atm	7 atm	10 atm	13 atm
At rest	−3·2	−6·9	−24·6	−61·6
During exercise	−2·1	−11·6	−39·8	—

noted that the narcosis is worse if the compression is rapid rather than slow. The effect of rate of compression is considered more fully in Chapter V.

That nitrogen may produce some degree of narcosis under normal or even reduced pressures was proposed by Miles (1962) who called this hypothesis the "nitrogen blanket theory". Recent experimental work does not however support this idea. Seventy subjects were exposed by Dunn (1962) to a multi-dimensional pursuit task whilst breathing various oxygen–nitrogen mixtures at atmospheric pressure. The partial pressure of nitrogen was reduced either by increasing the oxygen percentage whilst maintaining constant total pressure or by decreasing the total pressure at a constant oxygen partial pressure. No significant change could be correlated with the various nitrogen partial pressures but increases in the oxygen partial pressure decreased the magnitude of performance decrement. Similarly Hall and Kelly (1962) found no significant evidence for a "nitrogen blanket" in men exposed to 100 per cent oxygen at a simulated altitude of 34,500 feet for 5 days.

That compressed air does possess narcotic properties is readily apparent on consideration of the above work but the diversity of tests and the different results obtained by different workers even at the same nitrogen partial pressure make a clear understanding of the extent of compressed air on performance difficult.

In summary the minimum pressure at which the signs and symptoms of compressed air narcosis occur is disputable. Indeed it has been recently suggested that a decrement in performance at an unpractised task will occur at pressures as low as 2 atmospheres absolute of air (Poulton, Carpenter and Catton 1963). A considerable difference indeed to the 7 atmospheres absolute for practised tasks quoted by Miles and Mackay (1959).

What is less in dispute is that the signs and symptoms do become more severe with increasing depth or pressure. Further that reducing the oxygen partial pressure and increasing the nitrogen partial pressure at a constant total pressure enhances the narcosis. Similarly, increasing the oxygen partial pressure at a constant nitrogen partial pressure increases the narcosis, as does hard work and a relatively fast compression.

Intellectual functions are the most affected as is short term memory. Manual dexterity is affected to a lesser extent. There is a considerable interindividual susceptibility to the narcosis, whose onset is rapid and to which frequent exposure produces some acclimatization. Recovery on decompression is rapid.

Considerable care is required in carrying out psychometric studies of the kind described in this chapter and the careful control of the many gas variables cannot be over-emphasized.

The degree of motivation between different individuals participating in psychometric tests is a variable factor which is also difficult to control and the extent of learning and acclimatization to the effects of pressure are conditions which are often ignored. Effective control of the subjects, which have varied between trained and inexperienced divers, officers and men of the armed forces, medical students and the experimenters themselves is commonly ignored in spite of the knowledge that work, fatigue, apprehension and alcohol affect the severity of the narcosis. This is often because the psychometric tests are *ad hoc* investigations in experiments primarily designed to provide physiological information.

The need for a simple effective test of narcosis which can be widely used as an index of narcosis is an urgent need for the future. It would enable one to obtain a clear understanding of the interindividual variation to depth intoxication and assist in the selection of men best suited to deep diving or enable a warning to be given to the sensitive individual besides helping to increase our knowledge of the cause and mechanisms of the narcosis.

Other Inert Gases

Many other inert gases exert similar signs and symptoms of narcosis when breathed by man and animals as does the nitrogen content of air. Ether and nitrous oxide for example are well known as anaesthetics. Of particular interest however is that the so-called noble gases, radon, xenon, krypton, argon, neon and helium, at pressures depending on the gas, display all the properties characteristic of narcotics and anaesthetics (Rinfret and Doebbler 1961).

Signs and symptoms of narcosis are minimal with helium oxygen (End 1938, Behnke and Yarbrough 1938). The latter authors noted that at 500 feet the subjective narcosis was equivalent to some 100 feet of air. Little is known of the narcotic

TABLE 6. ESTIMATES OF DEPTH MADE BY MEN BREATHING ARGON–OXYGEN ON THEIR EXPERIENCE BREATHING AIR AT RAISED PRESSURES (BEHNKE AND YARBROUGH 1939)

Subject	Actual depth (ft)	Depth estimate
1	130	200
2	90	150
3	130	200
4	120	250

properties of neon but work at present being carried out at the R.N. Physiological Laboratory and by Captain George Bond, M.C., U.S.N. at the Naval Research Laboratories, New London, U.S.A., should shortly increase our knowledge of the effects of this gas. The narcotic potency of argon is greater than nitrogen. Divers breathing argon–oxygen believed, from their subjective experience of the narcotic effect of air, that they were almost twice as deep as they actually were (Table 6). During inhalation of 50 per cent krypton–50 per cent oxygen at atmospheric pressure subjects have reported dizziness and voice distortion (Lawrence et al. 1946, Cullen and Gross 1951).

The most potent narcotic of the noble gases, excluding radon due to its radioactive properties, is xenon. Cullen and Gross (1951) were the first to use xenon–oxygen as a surgical anaesthetic. At concentrations of 80 per cent xenon–20 per cent oxygen an orchidectomy was performed on an 81-year-old man and ligation of the fallopian tubes in a 38-year-old woman. Light anaesthesia in the first plane third stage was attained. Since that time many workers have examined and reported this property of xenon (Pittinger *et al.* 1951; Cullen and Pittinger 1952; Pittinger *et al.* 1953; Morris, Knott and Pittinger 1955; Featherstone 1960). Indeed, as is readily apparent, Pittinger and his colleagues at the University of Iowa have devoted considerable time and effort to the study of xenon (Pittinger 1962).

Investigations of the effects of the noble gases on man are however relatively sparse. A considerable amount of our knowledge of the narcotic properties of the noble or rare gases has been obtained from animal experiments.

The Effects of Inert Gases on the Whole Animal

Behnke *et al.* (1935) are normally awarded the distinction of being the first to attribute depth intoxication in man to the increased partial pressure of nitrogen. However, on the basis of animal experiments, Meyer and Hopff (1923) had already concluded that nitrogen was a narcotic at high pressures. Salamanders and frogs lay flat on their backs at 90–100 atmospheres of nitrogen, either not moving or exhibiting uncontrolled shaking. Insects (*Blatta orientalis* and *Blatta germanica*) were also narcotized at 90 atmospheres of nitrogen. The insects did not regain consciousness until 10 minutes after return to normal air pressures. This was attributed to the high oxygen requirements of these insects, which could not be satisfied when the pressure was decreased.

Although these workers attributed the narcosis to the nitrogen concentration, it is possible that the reported effects are the result of pressure alone. Certainly these very high pressures are much in excess of those producing narcosis in animals and comparisons must be made with caution.

Nevertheless, on the basis of these and similar experiments on other substances by Meyer and Gottlieb-Billroth (1920), Meyer proposed the following rule: "All gaseous or volatile substances will induce narcosis if they penetrate the cell lipids in a definite molar concentration, which is characteristic for each type of animal (or better type of cell) and is approximately the same for all narcotics." In the case of nitrogen, the molar concentration in amphibians was calculated as 0·18 mole/litre of lipid. For mice Carpenter (1954) calculated the molar concentration as 0·052 and in rats Bennett (1963) found the value was 0·033 mole/litre.

Although Cullen and Gross (1951) were the first to observe that xenon was an efficient human anaesthetic, Lawrence *et al.* (1946) appear to be the first to report the narcotic property of xenon. Five mice were exposed to various oxygen–xenon concentrations over periods up to a maximum of 1 hour. Within 2 minutes the animals showed convulsive extensor movements of the head and weakness or paralysis of the hind limbs.

Lazarev had inferred in 1941 that xenon might be narcotic at atmospheric pressure. This inference was based on the knowledge that the narcotic effect of the rare gases increases with their molecular weight. Studies of the effect of helium on adult mice showed no marked narcosis at 100 atmospheres, whereas argon was narcotic at 16–18 atmospheres. An immature mouse was narcotic at only 11 atmospheres. With krypton the same mouse was narcotic at $3\frac{1}{2}$ atmospheres.

Lazarev had to interrupt his experiments due to the war. However, Lawrence *et al.* were also working independently along similar lines in the United States as already discussed.

Lazarev *et al.* (1948) demonstrated that 48·5 per cent of xenon, with some krypton present, produced unconsciousness in an immature mouse. It was found that the narcosis was apparent in 3 minutes and that recovery was similarly rapid. An adult mouse in 75 per cent xenon was slow in recovering its normal position when the pressure chamber was turned upside down. The effect of xenon was also investigated on German cockroaches. At 3·1 atmospheres the insects could still move but were unsteady and lethargic. At a lower partial pressure of 2–2·5 atmospheres the insects were excited and moved rigorously around the pressure chamber. With a still lower pressure of 0·7–1·5 atmospheres the insects were torpid and remained motionless. On the basis of these experiments, Lazarev concluded that xenon had a narcotic effect rather more strong than krypton.

C

Extensive investigations of the effects of inert gases on frogs have been made by Marshall (1951) and Marshall and Fenn (1950). With adequate oxygen present, 41–61 atmospheres of nitrogen slowed the respiration, relaxed the legs and produced a dropping of the head. The response to an induction shock, normally a vigorous jump, was weakened. Helium at 68 atmospheres for 4 hours had no effect.

Mice were found to be more sensitive than amphibians (Marshall 1951). At 10–17 atmospheres of nitrogen they lost their sense of equilibrium in 15 minutes. Higher pressures resulted in staggering and paralysis. Helium did show some evidence of narcosis in mice. Fifty-four atmospheres caused decreased activity and staggering. During decompression the animals recovered completely.

Clearly inert gas narcosis is not restricted to mammals. Insects and amphibians are equally sensitive. Indeed Sears and Gittleson (1961) have shown that even unicellular organisms such as Paramecia are affected. At 210 lb/in^2 xenon the contractile vacuoles cease to function, movement stops and the organism swells.

It is apparent that whatever the cause and mechanisms responsible for the narcosis they are of fundamental origin. Some of the possibilities considered are discussed in the chapters that follow.

CHAPTER II

THE CAUSES OF THE NARCOSIS

EXPOSURE of man and animals to increased pressures of air or mixtures of the inert gases and oxygen results in a form of narcosis defined by the signs and symptoms and decrement in performance discussed in Chapter I. At first glance the cause of the narcosis appears relatively simple. It could be the result of either the increased nitrogen or inert gas partial pressure or the increased oxygen partial pressure. However, closer inspection reveals other possibilities which must be considered. Is the pressure itself involved? Are the effects in man related to the psychological characteristics of an individual? What part is played by the fact that the increased density of the gas mixtures breathed at increased pressures may make respiration difficult, especially if work is involved?

Many causes have in fact been suggested for the narcosis. Most of these lack experimental support and are readily disproved. The more common hypotheses will be considered in this chapter. It will soon be evident that the cause of the narcosis is not simple and that more than one factor is involved as suggested by the title of this chapter.

Pressure Alone

One of the earliest hypotheses for the cause of the narcosis was by Moxon who suggested in 1881 that the narcosis was a function of the pressure alone. The hypothesis implied that the increased pressure caused the blood to pool in the brain and spinal cord. The result was an increased intracranial pressure, which was responsible for the interference with the function of the central nervous system. There is however no blanching of the skin during the narcosis which might be expected if this theory were correct. The hypothesis is readily disproved by the fact that at the same pressure different inert gases cause different levels of narcosis.

As discussed more fully in Chapter I, xenon is an anaesthetic at atmospheric pressure. At this pressure krypton causes slight dizziness and argon, nitrogen and helium have no effect.

Psychological Origin

In the 1933 Admiralty Report on Deep Diving, considerable importance was given to the signs and symptoms being produced or exaggerated by psychological or physiological differences in the divers. A number of physiological tests were carried out, including investigation of pulse rate, blood pressure, vital capacity,

TABLE 7. BREATH-HOLDING TEST AS AN INDICATION OF SENSITIVITY TO DEPTH INTOXICATION (HILL et al. 1933)

Diver No.	Dives	Satisfactory	Dizziness, etc.	Breath-holding (sec)
12	3	0	3	38
1	4	1	3	32
2	4	1	3	49
11	4	2	2	61
9	4	2	2	45
4	5	3	2	55
8	6	4	2	54
5	4	3	1	72
10	5	5	0	102
6	5	5	0	54
7	6	6	0	59

pulse response test, 40 mm effort test, E.C.G. and breath-holding, together with a neurological examination. All the divers who reported signs and symptoms of narcosis had some physical imperfection and it was suggested that a thorough physical examination was a vital necessity for a man contemplating deep diving. However it was clear from the physiological tests that physical health was not the entire cause.

Accordingly, Dr. Millais Culpin, Psychological Adviser to the Medical Research Council, sought to ascertain if the divers possessed any psychological characteristics which might be responsible for the narcosis. Of 6 divers it was considered that at least 2 were really in a state of fugue or terror during the periods reported as semiconscious. It was suggested that if a diver was in such a state of terror, due perhaps to claustrophobia, the fact that his labyrinthine reaction was bad or he had other physiological defects in the upper respiratory tract, could exaggerate the psychological effect and initiate loss of consciousness. A breath-holding test on the divers showed some correlation with the liability of an individual to lose consciousness at depth (Table 7). The most unsatisfactory divers as regards narcosis were numbers 12, 1, 2, 11 and 9 and on average their breath-holding time was shorter than normal.

Besides being a physiological test this is also to some extent a test of willpower and may therefore have some relation to the psychological considerations. Although the fact that at the same pressure different inert gases exert different levels of narcosis is not in support of the psychological theory, there is no doubt that any deep diver must be physically fit and of the correct psychological temperament. In this connection the breath-holding test is a simple and useful guide to the aptitude of an individual.

The Raised Partial Pressure of Oxygen

That the narcosis might be due to the raised partial pressure of oxygen was postulated by Damant (1930). It is well known that oxygen at high pressures causes convulsions (Bean 1945) but that it is also the cause of the narcosis may be quickly disproved. Air at 10 atmospheres absolute (300 feet) has an oxygen partial pressure of 2 atmospheres absolute. Since 100 per cent oxygen at 2 atmospheres absolute produces none of the signs and symptoms of narcosis reported at 300 feet when breathing air, oxygen cannot be the cause of the condition. That it may exert a synergistic action due to its effect on carbon dioxide retention is considered in greater detail in Chapter III.

The Raised Partial Pressure of Inert Gas

(a) *The Narcotic Potency of Inert Gases*

In 1935 Behnke, Thomson and Motley were the first to attribute depth intoxication to the raised partial pressure of nitrogen in the air. This conclusion they based on the Meyer–Overton theory (1899, 1901) that there is a parallel between the affinity of an aliphatic anaesthetic for lipid and its narcotic or depressant action. The more soluble an agent is in lipid and the less soluble in water, the more potent it will be as an anaesthetic. Since nerve cells are rich in lipid, the narcotic is thought to gain access to nerve tissue by virtue of its lipid solubility. Support for this theory rests on the fact that arrangement of anaesthetic agents in order of narcotic potency agrees reasonably with the order found by grouping them according to their solubility coefficients (Behnke and Yarbrough 1939). Examination of the oil solubilities in Table 8 illustrates this correlation. A number of anomalies are however apparent.

As illustrated in Table 6, Chapter I, argon is approximately twice as narcotic as nitrogen, which is not in agreement with the partition coefficients. However argon is twice as soluble as nitrogen in both oil and water. An anomaly that is not so readily explained is that in spite of the high relative oil solubility of hydrogen, it is of low narcotic potency. Case and Haldane (1941) demonstrated that the narcotic effect of air at 8·6 atmospheres absolute (250 feet) is abolished by substitution of an oxygen-hydrogen mixture and Zetterstrom (1948, 1949) was compressed to 16·7 atmospheres absolute (535 feet) on a 4 per cent oxygen-hydrogen mixture, without noticing any signs and symptoms of narcosis. Indeed, there are workers who believe that hydrogen is less narcotic than helium. However as the partition coefficient and oil solubility for helium is lower than hydrogen this would seem most unlikely. It is perhaps unfortunate that hydrogen has been included in considerations of this nature, for in many ways it is a relatively active gas. Due to the explosive nature of oxygen-hydrogen mixtures few investigations have been made of its narcotic nature. Under the circumstances it might well be better if it were not included in future considerations of the aetiology of depth intoxication, certainly until more data are available.

Oil solubility values for neon are not available at present† and the values for the water solubility of this gas need confirmation. However investigations at the Medical Research Laboratories, New London, U.S.A. may shortly rectify this gap in our knowledge.

There are many other anomalies and exceptions to the Meyer-Overton theory. Many substances with a high oil solubility do not possess narcotic properties and vice versa, e.g. chloral hydrate. This is perhaps not surprising considering the theory relies

---

† The oil solubility values quoted in Tables 8 and 11 have been inserted as a result of recent work by Ikels (1964). See Addendum, Chapter VIII.

TABLE 8. PARTITION COEFFICIENTS OF THE INERT GASES

	Helium	Neon	Hydrogen	Nitrogen	Argon	Krypton	Xenon
Oil/water solubility ratio (37°C)	1·7:1	2·07:1	3:1	5·2:1	5·3:1	9·6:1	20:1
Narcotic potency	Mildly narcotic			⟶			Anaesthetic

TABLE 9. ACTIVITIES DERIVED FOR ANAESTHESIA IN MAN (BRINK AND POSTERNAK 1948)

Gas	Volumes % in inspired gas	Partial pressure in H$_2$O-saturated gas at 37°C	Vapour pressure of pure narcotic	$A_{nar} \times 10^2$ (Activity)
Ether	4	29 mm Hg	837	3·4
Chloroform	1·5	11 mm Hg	322	3·3

mainly on *in vitro* measurements of solubility in olive oil or alternatively corn oil, cottonseed oil and paraffin oil. The lipid phase in man has very different characteristics from these.

Nevertheless, the fact remains that here is an analogy which does seem to have a relation with the narcotic properties of the inert gases. Another analogy in physico-chemical terms has been prepared by Ferguson (1939) and is based on the thermodynamic activity of the inert gases.

The concept of chemical potentials as indices of toxicity relies on the assumption that, in cases of physical toxicity, there exists a concentration equilibrium between the external environment and the concentration of the agent responsible for narcosis in the animal. The activity function is related to this concentration but it is corrected for deviations from that of the ideal gas. Ferguson demonstrated that for a variety of narcotics their vapour concentrations, expressed as *activity*, are a better indication of isonarcotic potency than volumes per cent or moles per litre.

Brink and Posternak (1948) supported this concept (Table 9) but indicated from experiments on the cat stellate ganglion that this rule did not apply to non-synaptic paths in the nervous system but only to the synaptic pathways (Table 10). The biological system under examination must therefore be considered if isonarcotic potency is to be expressed in the form of activity.

TABLE 10. REVERSIBLE SUPPRESSION OF TRANSMISSION OF NERVE IMPULSES
(BRINK AND POSTERNAK 1948)

Molecule	Synaptic pathways narcotic concn. mole/l	$A_{nar} \times 10^2$	Non-synaptic pathways narcotic concn. mole/l	$A_{nar} \times 10^2$
n-Ethanol	0·478	3·2	0·240	1·6
n-Butanol	0·048	4·6	0·048	4·6
2-Hexanol	0·00314	2·5	0·00668	5·3
n-Octanol	0·00012	2·7	0·00055	12·0
Chloroform	0·0044	6·3	0·0132	19·0
Ethyl ether	0·0336	4·2	0·096	12·0

A further analogy between narcotic potency of the inert gases and other narcotics and their Van der Waals physico-chemical constants has been made by Wulf and Featherstone (1957). Increasing potency of clinical anaesthetics was shown to compare favourably with the increasing magnitude of these constants

(Table 11). The authors stressed that this inferred that molecular volume (constant "b") was an important factor in the mechanism of the narcosis. Again however there are anomalies, such as the inference that argon may be less narcotic than nitrogen and the

TABLE 11. RELATIONSHIP BETWEEN NARCOTIC POTENCY AND VAN DER WAALS CONSTANTS

	(1) Constant (a)	(2) Constant (b)	(3) Oil solubility 37°C	(4) (1)×(3)	(5) (2)×(3)
Neon	0·2107	1·709	0·019	0·004	0·03
Helium	0·0341	2·370	0·015	0·0005	0·04
Hydrogen	0·2444	2·661	0·04	0·01	0·11
Argon	1·345	3·219	0·14	0·19	0·45
Nitrogen	1·390	3·913	0·067	0·09	0·26
Krypton	2·318	3·978	0·043	1·0	1·7
Xenon	4·194	5·105	1·7	7·1	8·68

very low position of neon. Marshall (1950) has shown that neon is less narcotic than nitrogen but more narcotic than helium. Elsewhere however Marshall (1951) supported the relationship between inert gas narcosis and thermodynamic activity. At the concentrations required to depress frog brain waves the thermodynamic activity (A_{nar} values) for nitrogen and argon were 0·0517 and 0·0461 and the values for inhibition of the frog reflex were 0·0165 and 0·0119 respectively (Table 12).

TABLE 12. COMPARISON OF THRESHOLD NARCOTIC CONCENTRATIONS OF THE GASES ON THE BASIS OF THEIR LIPID SOLUBILITY AND THERMODYNAMIC ACTIVITY (MARSHALL 1951)

Gas	Mol. wt.	Thresh. press. atm	f_0 atm	f atm	$A_{nar} \times 10^2$	Sol. cc/cc oliveoil.22°C	C_{lip} mole/l
\multicolumn{8}{c}{Inhibition of Brain Waves}
N_2	28	54	1030	53·4	5·17	0·067	0·112
A	40	41	860	39·8	4·61	0·140	0·250
\multicolumn{8}{c}{Inhibition of Reflex Activity}
N_2	28	17	1030	17	1·65	0·067	0·036
A	40	10	860	10	1·19	0·140	0·063

f_0 = Fugacity of narcotic at its vapour pressure at 25°C
f = Fugacity of narcotic concentration in experiment
A_{nar} = Thermodynamic activity of narcotic f/f_0

It has been suggested by many workers that there is a relationship between the potency of an inert gas as a narcotic and its oil solubility. Both Wulf and Featherstone (1957) and Sears (1962) have further suggested that the inert gases produce the narcosis by an effect on the lateral spacing of the lipid molecules causing an interference with the permeability of ions across cell membranes.

A more correct order of narcotic potency might therefore be afforded if one considers that oil solubility is equally as important as molecular volume. Reference to Table 11 shows this to be true and comparative narcotic relationships more in keeping with our general knowledge of the narcotic properties of these gases are obtained.

The Van der Waals constant "a", which is a measure of the attractive forces of molecules for one another, when multiplied by the oil solubility does not give such a good correlation although the relationship is better than that based on constant "a" alone. It would seem that molecular volume is of greater importance in the mechanism of narcosis than molecular attraction.

Recently Featherstone and Muehlbaecher (1963) have stressed that the majority of the physical properties of substances correlated with narcotic potency are simply reflections of their intermolecular Van der Waals attractions.

Among such physical properties considered are molecular weight (Table 13, Behnke and Yarbrough 1939), adsorption (Case and Haldane 1941), oil solubility (Meyer–Overton 1899, 1901), thermodynamic activity (Ferguson 1939, Brink and Posternak 1948) and the formation of clathrates (Miller 1961, Pauling 1961).

A well-voiced criticism of the inert gas partial pressure as the cause of depth intoxication is that the greatest mental change is usually evident immediately on reaching pressure or even during compression and becomes less on remaining at pressure (Case and Haldane 1941, Burnett 1955, Unsworth 1960). It might reasonably be expected that as the nitrogen or other inert gas tension increases with the increased time of exposure, then the narcosis should become more severe with time at pressure.

The results of psychometric tests do not however support any correlation between the level of narcosis and time at pressure (Kiessling and Maag 1962). It is likely that the increased subjective narcosis is due to a combination of the psychological stress of compression, undue temperature changes, the noise of compression and the use of the valsava manoeuvre to "clear the ears".

TABLE 13. MOLECULAR WEIGHTS OF INERT GASES AND NARCOTIC POTENCY

Gas	Hydrogen	Helium	Neon	Nitrogen	Argon	Krypton	Xenon
Molecular weight	2	4	20	28	40	83·7	131·3
Narcotic potency	Mildly narcotic	───────────────────────────→					Anaesthetic

A further factor, which is considered in greater detail in the next chapter, is the change in cerebral carbon dioxide tension due to compression.

(b) *Inert Gas Saturation*

There is little reason why symptoms should increase with exposure. Although saturation of the whole body may take some time, the highly vascular central nervous system reaches equilibrium in a much shorter time. Pittinger *et al.* (1956) examined the kinetics of transfer of radioactive xenon and chloroform between blood and brain in dogs. Their study showed that the transfer was a flow-limited process and Ferris, Molle and Ryder (1942) showed that exchange of nitrogen with arterial blood was extremely rapid. One passage of the blood through the lungs is enough to empty or fill it with nitrogen to within 80–95 per cent of the equilibrium value. Saturation in these circumstances should be rapid.

Pittinger *et al.* (1954) confirmed that cerebral saturation was indeed rapid, even with a dense gas such as xenon. The thalamus,

TABLE 14. XENON CONTENT IN MICROLITRES PER MILLIGRAM OF TISSUE (DRY WEIGHT) (PITTINGER *et al.* 1954)

	Period of inhalation		
	2 min	6 min	20 min
Cerebral parietal cortex	0·28	0·33	0·44
Caudate nucleus	0·41	0·40	0·42
Thalamus	0·41	0·45	0·44
Hypothalamus	0·36	0·40	0·44
Medulla oblongata	0·40	0·44	0·40
Adrenal gland	0·34	0·41	0·67
Kidney	0·28	0·31	0·38
Liver	0·18	0·23	0·26
Striated muscle	0·03	0·03	0·14

hypothalamus, caudate nucleus and medulla oblongata are almost fully saturated within 6 minutes and the cerebral cortex probably within 20 minutes (Table 14).

Exposure to compressed air at 14·2 atmospheres absolute (450 feet), by sudden removal of a mask giving helium–oxygen, resulted (Wood 1962) in a marked decrease in all psychometric parameters beginning as early as 30 seconds after starting to breathe air. Within 2½ minutes all subjects showed a performance decrement of such an extent they would have been in considerable danger if they had been diving in water instead of carrying out a simulated dive in a pressure chamber.

Although these facts and the previously discussed analogies between the narcotic potency of the inert gases and their physical characteristics do suggest that the inert gases are responsible in themselves for producing the narcosis, they do not constitute proof. Such proof is however available and is discussed in the following chapter, which considers an alternative theory that the narcosis is due to retained cerebral carbon dioxide. Further evidence is also considered in the chapter concerned with electrophysiological investigations of the problem.

CHAPTER III

TISSUE CARBON DIOXIDE RETENTION AS THE CAUSE OF INERT GAS NARCOSIS

CONSIDERATION of Chapter II suggests that the cause of the signs and symptoms of depth intoxication and inert gas narcosis is the result of the increased partial pressure of nitrogen or inert gas in the mixture breathed. There is however another possibility which merits consideration in some detail. This is that the narcosis is due to an increased carbon dioxide tension in the tissues as a result of lung ventilation difficulties.

Although Hill and Greenwood (1906) had discounted the possibility, the Admiralty Deep Diving Committee in 1933 again raised the possibility that the narcosis was due to carbon dioxide retention but became more interested in the possibility of psychological involvement. Carbon dioxide was not seriously considered again until recently by Bean (1947, 1950). There is now an increasing number of workers who suggest that the signs and symptoms of depth intoxication could be the result of carbon dioxide narcosis. This hypothesis is considered in this chapter.

Alveolar Carbon Dioxide

During experiments to determine the changes in the pH of arterial blood throughout compression and decompression of anaesthetized, heparinized dogs, Bean (1947) noted a change in the acid direction (0·15–0·02 pH) during compression and the reverse during decompression. It was inferred that the cause was due to an interference with removal of carbon dioxide.

This could be affected by the compression of the alveolar gases producing an increase in the partial pressure of CO_2 together with the inrush of air into the lungs during rapid compression preventing CO_2 output and the difficulty of diffusion and mixing the CO_2 meets due to the raised density of the air. Experiments on 24 anaesthetized dogs further substantiated the theory (Bean 1950). Alveolar samples were obtained through a catheter draining the lung periphery at atmospheric and increased air pressures of 4·7, 6·2 and 6·9 atmospheres (70, 90 and 100 lb/in²). These showed an increase in alveolar CO_2 (Fig. 3) from the normal

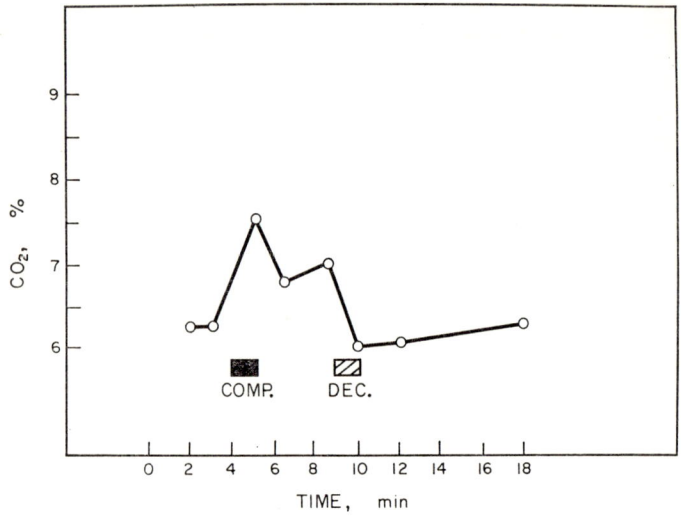

FIG. 3. Alveolar pCO_2 of anaesthetized dog during compression and decompression to 90 lb/in² (Bean 1950).

5 per cent to values as high as 10 per cent of 1 atmosphere in one case. The increase in alveolar CO_2 was especially prevalent if the compression was rapid. Bean concluded that until definite proof otherwise was available, carbon dioxide should be considered as an alternative cause of the narcosis.

In an effort to obtain this proof Rashbass (1955) exposed 26 men to a pressure of 8·6 atmospheres absolute of air (250 feet). Measurement of the degree of narcosis was made by the use of two-figure by one-figure multiplication, the score being the number of correct answers achieved within 2 minutes. Alveolar gas samples were collected by the method of Haldane and Priestly (1935). Tests were given and samples taken at atmospheric pressure, immediately on reaching total pressure, after 5 minutes' hyperventilation and after 5 minutes without hyperventilation. The results (Table 15) led Rashbass to conclude that the suggestion that carbon dioxide is responsible for causing the narcosis was fairly adequately disproved. The extent to which carbon dioxide contributed to the rapid onset or increased initial narcosis remained undecided.

TABLE 15. ALVEOLAR pCO_2 AT HIGH AIR PRESSURES (RASHBASS 1955)

	Mean alveolar $CO_2\%$ atm	Mean sums correct
Surface	5·125	24·12
250 ft	5·385	16·81
250 ft after H./V	3·91	18·0
250 ft without H./V	5·21	18·2

That carbon dioxide is not the cause of the narcosis is supported by the results of similar experiments by Cabarrou (1959, 1964). Alveolar CO_2 was measured by a suction technique, together with measurements of respiratory output at air pressures of 5, 5·7 and 9 atmospheres. Cabarrou observed an increase in the alveolar CO_2 to 7·5 per cent of an atmosphere at 5 atmospheres. This returned to 5·3–5·4 per cent within 4–8 minutes. If compression was very slow there was no change in the alveolar CO_2. Hyperventilation before compression also ensured no increased alveolar CO_2 at pressure.

At 9 atmospheres, with a rapid compression, the alveolar pCO_2 rose to 6–8 per cent and in one case 10 per cent of an atmosphere and then returned to normal. Correct ventilation of the lungs at pressure reduced the partial pressure of CO_2 to practically normal levels.

A further indirect proof against the CO_2 theory was made by exposing subjects to mixtures containing 2 per cent oxygen–98 per cent nitrogen at 9 atmospheres. The partial pressure of oxygen was now little different to that of air at atmospheric pressure. The same subjects were also compressed to 7 atmospheres with a mixture of 2·5 per cent oxygen–97·5 per cent nitrogen, again giving an almost atmospheric oxygen partial pressure. An excess of CO_2 in the brain under these conditions should have caused an immediate increase in the rate and amplitude of ventilation but no variation was observed during 60 experiments.

Examination of the alveolar CO_2 measurements of Bean (Fig. 3) show that the results are very similar to those of Cabarrou. The latter however made measurements at periods of 1, 2, 4, 6, 8, 10, 15, 20 and 30 minutes at pressure whereas Bean recorded every 1–2 minutes for only some 3–6 minutes at pressure. Most of the graphs by Bean as illustrated in Fig. 3 show a gradual fall in the carbon dioxide a few minutes after reaching total pressure. Had recordings been made over a longer period it is likely that Bean would also have observed a return to normal values after 6–8 minutes.

Recently Buhlman (1961, 1963), Seusing and Drube (1960) and Seusing (1961) again denied the narcotic action of nitrogen and proposed carbon dioxide as the cause, using much the same argument as Bean. This suggests that due to the increased gas density caused by compression, there is a rise in the viscous resistance in the respiratory tree. This increased breathing resistance causes hypoventilation, with the resulting retention of carbon dioxide. The problem is aggravated by the high partial pressure of oxygen which also produces a tendency to hypoventilation. In this manner there arises a hyperoxic hypercapnia. It is suggested that the narcosis may be avoided by using a less dense gas such as helium and oxygen or by using auxiliary breathing apparatus to ensure satisfactory alveolar ventilation.

Buhlman has used both assisted ventilation and helium to overcome the narcosis and Hans Keller, his diving colleague, is reported to have attained a pressure of some 11·9 atmospheres absolute (360 feet) without being narcotic whilst breathing a mixture of oxygen–nitrogen which was low in oxygen. As this would mean that the equivalent air depth, based on the nitrogen partial pressure, would be more than 400 feet, this is difficult to understand. However, subjective appreciation of the narcosis is very unreliable. Objective experiments do not support the inference that reduction of oxygen and increase thereby of the inert gas partial pressure reduces the severity of inert gas narcosis. Indeed the reverse would appear to be the case (Chapter I, Table 3, Albano *et al*. 1962). Other electrophysiological experiments of a similar nature as those described by Albano *et al*. (1962) also do not support Buhlman.

Electrophysiological Studies, Density and Breathing Resistance

Bennett, Dossett and Kidd (1960) exposed rats to various oxygen–nitrogen partial pressures at the same total pressure using an electroshock technique (Bennett 1963) as the index of narcosis. It may be seen from Table 16 that the percentage rise in volts

TABLE 16. EFFECT OF DENSITY AND OXYGEN PARTIAL PRESSURE ON NARCOSIS (BENNETT, DOSSETT AND KIDD 1960)

Mixture	Equivalent density	% rise in V for stimulus
5 O_2–190 N_2 lb/in^2	5480	120
25 O_2–170 N_2 lb/in^2	5660	48
45 O_2–150 N_2 lb/in^2	5460	20
65 O_2–130 N_2 lb/in^2	5720	15

required after compression to give the same animal response at atmospheric pressure is related to the nitrogen partial pressure rather than to the density. Decreasing the oxygen partial pressure did not lower the level of narcosis. Indeed the narcosis was worse. Similar findings have been reported in man by Barnard, Hempleman and Trotter (1962) (Chapter I, Fig. 1) and by Albano and Ciulla (1962) in studies of respiratory minute volume at different oxygen–nitrogen partial pressures.

That the narcosis will however be facilitated if the oxygen partial pressure is increased at a constant partial pressure of nitrogen is considered in more detail in Chapter I.

In a recent personal communication Buhlman indicates that he interprets the term narcosis as loss of consciousness. This suggests that the primary difficulty is one of semantics. For such an interpretation is in excess of that understood by the majority of workers in this field as is discussed in the Introduction and Chapter I.

There is no doubt that increased breathing resistance does occur at pressure as illustrated in Table 17 (Wood, Leve and Workman 1962). Although there remains sufficient ventilation capacity for mild activity, paradoxically the respiratory effort is not sufficient for adequate alveolar ventilation. The result is an increase in the alveolar CO_2 to often very high levels. However, as discussed later and illustrated in Figs. 4 and 5, the increased CO_2 may not necessarily produce signs and symptoms of narcosis unless there is an associated increased inert gas partial pressure. Buhlman and

Seusing, in favour of the carbon dioxide theory, further suggest that many workers in favour of the nitrogen theory have not considered that it is possible to cause the signs and symptoms of narcosis with an inert gas heavier than nitrogen such as xenon or krypton at atmospheric pressure. This, Buhlman suggests, proves the importance of molecular weight of the inert gases in a respirable mixture.

TABLE 17. MEAN DECREASE IN MAXIMUM BREATHING CAPACITY OF 11 SUBJECTS AT PRESSURES OF 1–15 ATM ABS. (WOOD, LEVE AND WORKMAN 1962)

Depth	Air		Helium	
	% decrease in M.B.C.	% of predicted M.B.C.	% decrease in M.B.C.	% of predicted M.B.C.
1	—	135·1	—	171·5
2	22·5	99·3	18·8	132·2
3	36·5	81·0	29·5	114·8
4	43·8	71·8	36·6	103·5
6	55·8	60·1	44·2	95·8
9	60·6	51·1	53·6	78·8
15	75·4	34·8	66·8	59·0

In fact, as described in the previous chapters, these factors have been appreciated for many years. Perhaps of still greater importance is that the density of a 20:80 oxygen–xenon mixture at 1 atmosphere is 111·44 and the calculated density of air at 10 atmospheres (300 feet) is 288. This is over twice the density of the xenon mixture but although there is evidence of narcosis if air is breathed at 10 atmospheres, one is not anaesthetized to the surgical levels found with xenon mixtures at atmospheric pressure (Cullen and Gross 1951, Morris, Knott and Pittinger 1955).

Extensive experiments with 26 subjects, in which the electroencephalogram (E.E.G.) was used to record changes in brain wave activity at pressure also do not support the carbon dioxide theory (Cabarrou 1959, 1964). A comparison was made between the effect of the partial pressure of oxygen, the effect of the partial pressure of nitrogen, the effect of the partial pressure of nitrogen at constant density and the effect of density variations at a constant partial pressure of nitrogen. Subjective and objective changes, as a result of the narcosis, correlated well with changes in the E.E.G. Others have also found a reliable correlation with the E.E.G. (Morris *et al.* 1955, Bennett and Glass 1961).

The results of these investigations showed that with a constant partial pressure of nitrogen, variations in the absolute pressure had no effect on the E.E.G., whereas with a constant absolute pressure variations in the partial pressure of nitrogen caused modifications to the E.E.G. Cabarrou concluded that the density of the mixture had no effect on the narcosis. He also concluded that carbon dioxide retention is not the cause of depth intoxication or inert gas narcosis except as an aggravating factor. The main cause of the narcosis is the raised partial pressure of the inert gas itself.

Blood pCO_2

Further clarification of the carbon dioxide versus nitrogen theories is afforded by examination of blood CO_2 levels. Morris, Knott and Pittinger (1955) measured the blood gases in human surgical patients during xenon anaesthesia (Table 18). Blood oxygen and carbon dioxide were found to be within normal

TABLE 18. BLOOD GAS ANALYSIS BEFORE AND DURING XENON ANAESTHESIA (MORRIS et al. 1955)

Subject	Before anaesthesia				After 15 min or more xenon anaesthesia			
	Arterial		Venous		Arterial		Venous	
	O_2 vol %	CO_2 vol %	O_2 vol %	CO_2 vol %	O_2 vol %	CO_2 vol %	O_2 vol %	CO_2 vol %
1	—	—	17·4	66	—	—	19·8	64
2	—	—	10·5	51	21·7	41	18·6	43
3	—	—	12·0	56·8	20·1	61·2	17·4	53·7
4	16·1	51·6	11·0	54·1	19·8	52·6	13·5	55·6
5	20·1	65·6	13·1	73·9	—	—	16·3	76
6	—	—	—	67·8	—	—	12·4	62·2
7	—	—	10·3	—	—	—	12·4	—

ranges which suggested efficient ventilation for removal of CO_2 and good oxygenation.

The exposure of rats and guinea pigs to 97 per cent nitrogen–3 per cent oxygen at 7 atmospheres absolute (200 feet) for 14 days (Workman, Bond and Mazzone 1962) also demonstrated no significant change in blood pH, pCO_2 or pO_2 (Table 19), although the rats seemed sluggish and tended to drag their hind legs.

TABLE 19. BLOOD STUDIES ON ANIMALS EXPOSED TO 97 PER CENT NITROGEN–3 PER CENT OXYGEN AT 7 ATM ABS. FOR 14 DAYS (WORKMAN, BOND AND MAZZONE 1962)

	Immediate		10 days survival		Control	
	Rats	G. Pigs	Rats	G. Pigs	Rats	G. Pigs
Mean pH	7·37	7·41	7·39	7·41	7·38	7·40
Mean CO_2 (vol %)	49·2	52·1	48·6	51·8	50·8	52·5
Mean O_2 (vol %)	11·6	12·4	11·8	12·4	12·1	12·8

Cortical pCO_2

As the narcosis is obviously due to an effect on the brain and as brain pCO_2 may vary independently of alveolar and blood pCO_2 levels (Dusser de Barenne et al. 1938, Merwarth and Sieker 1961, Meyer, Gotoh and Tazaki 1961) cerebral pCO_2 values are the only true indication of the role of CO_2 and the narcosis.

The cortical pCO_2 of cats has been measured at high pressures of inert gases by Bennett (1963b, 1965) using a modified Severinghaus blood pCO_2 electrode (1959). In Fig. 4 is shown the effect on the pCO_2 of exposing the animals to 130 lb/in² helium,

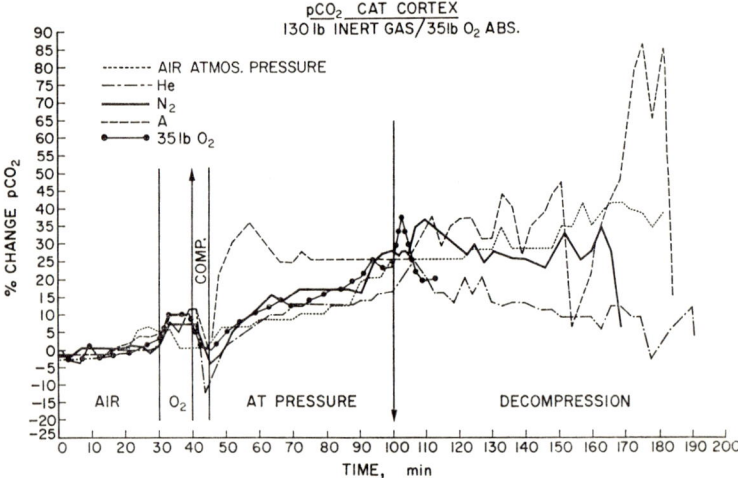

FIG. 4. Cortical pCO_2 of chloralosed cats exposed to 130 lb/in² argon, nitrogen and helium in the presence of 35 lb/in² oxygen, 35 lb/in² oxygen alone and at atmospheric pressure. Standard deviation for animals exposed to argon ±10·1 per cent, nitrogen ±16 per cent, helium ±7·7 per cent, air at atmospheric pressure 4·8 per cent, oxygen alone ±6 per cent (Bennett 1963b, 1965).

nitrogen or argon, in the presence of 35 lb/in^2 oxygen compared with a control at 35 lb/in^2 oxygen alone and another at atmospheric pressure. The effect of breathing 100 per cent oxygen at atmospheric pressure is also shown. The latter caused a rise of 7–10 per cent from normal values. During compression the cortical pCO_2 fell. This differs from the increased alveolar pCO_2 measurements of Cabarrou (1959, 1964) and Bean (1950). They inferred that the tissue pCO_2 was equally high but such is apparently not the case. The postulated cause of the increased initial narcosis is therefore unlikely to be due to an increased cortical pCO_2. At pressure the greatest increase in cerebral pCO_2 is found in animals breathing argon–oxygen. The nitrogen mixture shows a slight increase in pCO_2 but it is not significantly different from that found in cats breathing 35 lb/in^2 oxygen alone. For the helium–oxygen mixture the pCO_2 is not significantly different from control values. The latter tended to rise throughout the experiment, due to the animals being anaesthetized with chloralose (45–50 mg/g).

Measurements of frequent auditory evoked cortical potentials as an indication of the degree of narcosis were also made. The nitrogen mixture caused a 25 per cent reduction of the spike heights of the potentials, whereas the oxygen alone produced no effect. Under the circumstances the increased carbon dioxide tension cannot be responsible for the depression of the potentials and the narcosis.

Experiments where the oxygen partial pressure was not increased suggests in support of Hesser (1963) that the increase in pCO_2 is as much due to the high oxygen partial pressure as to the density of the mixture breathed (Fig. 5).

At the low oxygen partial pressure of 3 lb/in^2 the nitrogen mixture did not cause cortical pCO_2 values significantly different from the control. Narcosis was present however, as the auditory evoked cortical potentials were reduced by 35 per cent. The more dense argon mixture did cause some increase in cortical pCO_2, due no doubt to interference with the mechanics of respiration.

The hypocapneic values found with the oxygen–helium mixture are of considerable interest. In spite of the greater density of the oxygen–helium mixture compared with air at atmospheric pressure, the pCO_2 level is very much lower for the helium mixture.

Linaweaver (1961) has suggested that oxygen convulsions take longer to occur for a given oxygen partial pressure when breathing oxygen–helium compared with oxygen alone. When one considers the importance given to carbon dioxide as one of the causes of oxygen toxicity (Lambertsen 1955) this may not be

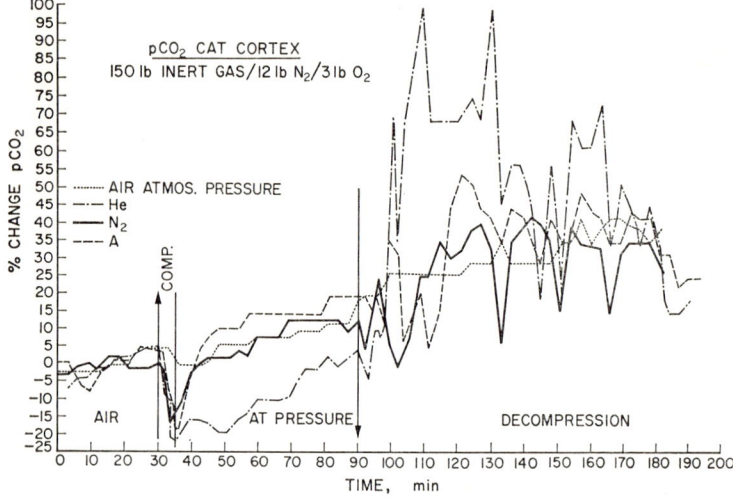

FIG. 5. Cortical pCO_2 of chloralosed cats exposed to 150 lb/in² argon, nitrogen and helium in the presence of air and air at atmospheric pressure. Standard deviation for animals exposed to argon ±7 per cent, nitrogen ±10·9 per cent, helium ±11·9 per cent, air at atmospheric pressure 4·8 per cent. (Bennett 1963b, 1965).

surprising. Recent unpublished experiments comparing the times to produce a convulsion in Wistar rats at 100 lb/in² oxygen and a mixture of 100 lb/in² oxygen–100 lb/in² helium suggest that there is no significant difference (Bennett 1964). This would suggest that the degree of pCO_2 change was not significant to affect oxygen toxicity. It is evident that there is much that is not known regarding the effect of helium breathing on tissue gas tensions and respiratory mechanics.

Symptoms such as paraesthesia and dizziness have been reported by men breathing low oxygen–high helium mixtures at equivalent depths of 400–500 feet. Are these due to inert gas narcosis, hypocapnia or the synergistic action of hypercapnia and inert gas narcosis? An answer must await the results of future experiments.

Synergistic Action of Carbon Dioxide and Oxygen

Although depth intoxication is therefore the direct result of the nature and partial pressure of the inert gas, increased oxygen partial pressures may exert a synergistic depressant action. This is due to the associated rise in cerebral pCO_2 as a result of the haemoglobin being fully saturated with oxygen. Oxyhaemoglobin circulates unchanged and this results in an increase in tissue pCO_2 (Gesell 1923).

Case and Haldane (1941) noted that whereas 3–4 per cent carbon dioxide at atmospheric pressure had no effect on manual or arithmetical skill, at 300 feet (10 atmospheres absolute) 3 per cent CO_2 caused a marked deterioration in manual dexterity and a state of confusion developed. Two subjects tested with 6 per cent CO_2 at atmospheric pressure were similarly unaffected but when breathing 6·6–9·7 per cent CO_2 at 300 feet (10 atmospheres absolute), eight subjects became unconscious within 5 minutes. Case and Haldane concluded that the combined effects of carbon dioxide and inert gases at high pressure were more severe than either alone. They suggested that in diving procedures the percentage of carbon dioxide in air at 10 atmospheres absolute should not exceed 0·3 per cent.

Further evidence of the synergistic action of carbon dioxide on inert gas narcosis was obtained by Marshall (1950). It was observed that 9·7 per cent CO_2 in the presence of increased partial pressures of nitrogen, argon and helium lowered the narcotic threshold for the inhibition of reflex activity. The thresholds were reduced from 17 atmospheres to 6·8 atmospheres for nitrogen, from 10 atmospheres to 6·8 atmospheres for argon and from more than 82 atmospheres to 20 atmospheres for helium.

Inert gas narcosis is therefore caused by a combination of factors. The principal cause is the nature and partial pressure of the inert gas. By virtue of their effect on carbon dioxide retention and its synergistic action the oxygen partial pressure and the density of the breathing mixture may also be pertinent factors.

CHAPTER IV

MECHANISMS OF THE NARCOSIS

Just as the dominating role of nitrogen or inert gases as the cause of the narcosis is not considered by some to be fully proved, leading to conflicting views and a number of hypotheses as to the cause, so with the mechanisms of the narcosis. This is perhaps not surprising, for the manner by which seemingly inert anaesthetics are able to produce unconsciousness still remains a mystery, in spite of volumes of research.

Do the inert gases interfere with nerve transmission and the complex polysynaptic connections within the brain, and if so, how? Is the narcosis the result of a tissue hypoxia at some specific site within nerve cells? In spite of their seeming inertness, is metabolism in fact affected by these gases? Can the inert gases form compounds after all? It is apparent that as with so many of the problems associated with the aetiology of inert gas narcosis there are many possible mechanisms.

Due to the inert nature of the rare gases and nitrogen, i.e. their inability to form covalent or hydrogen bonds in biochemical processes within living tissue, physical rather than chemical properties must be considered in endeavouring to explain the mechanisms by which they produce narcosis.

Pittinger and Keasling (1959) and Butler (1950) have already reviewed the many theories of narcosis and anaesthesia in relation to the clinical anaesthetics. This chapter is therefore devoted only to mechanisms which experiments suggest may be those by which the inert gases and compressed air cause narcosis, for it is unlikely that all narcotics and anaesthetics produce their effect by the same mechanism (Danielli 1950, Butler 1950).

Peripheral Nerve and Synaptic Transmission

It is a reasonable first assumption that the narcosis is due to interference at some level with the normal functions of the

MECHANISMS OF THE NARCOSIS 43

TABLE 20. COMPARATIVE NARCOTIC EFFECT OF 11 AGENTS INCLUDING THE INERT GASES AND NITROGEN RELATED TO NEUROPHYSIOLOGICAL MEASUREMENTS, ACTIVITY, OIL SOLUBILITY AND MOLAR CONCENTRATION (CARPENTER 1954)

Gas	$E.D._{50}$ atm	Olive oil Bunsen coeff. 37°	Olive oil Concn M/L.	$f°$ atm 37°C	A_{nar}	Fibre blockade atm	B.p. $E.D._{50}$	Δ demarcation potential at b.p.
Cyclopropane	0·045	7·15	0·036	7·5	0·0060	1·7–1·9	36	−10%
Dichlorodifluoromethane	0·26	5·1	0·057	7·0	0·037	3·4–4·7	18·5	−8%
Ethylene	0·47	1·28	0·026	49·0	0·0096	9·8–12·5	26	−12%
Xenon[1]	0·51	1·7	0·038	52·0	0·0098		not determined	
Nitrous oxide	0·58	1·6	0·036	44·0	0·013	10–13	23	−19%
Krypton[1]	1·8	0·43	0·034	215·0	0·0084		not determined	
Sulphur hexafluoride	1·87	0·25	0·020	20·4	0·091	≥21†	≥11	
Methane	2·9	0·28	0·043	262·0	0·011		36	−16%
Argon	12·6	0·14	0·077	725·0	0·017	92–110	27	−11%
Nitrogen	18·0	0·067	0·052	1700·0	0·010	310–340	not determined	
Helium	>163·0	0·015	>0·107			*		*

* Control at 350 atm: no observable change in action potential or demarcation potential.
† In excess of its vapour pressure at 25°C.
[1] Lazarev, N. (estimated $E.D._{50}$).

nervous system. However, it was not until as recently as 1950 that any proof was forthcoming. Marshall and Fenn (1950) observed that nitrogen at 17 atmospheres and argon at 10 atmospheres reversibly blocked the isolated frog reflex preparation after some 260 minutes. Helium at pressures as high as 82 atmospheres had no effect. On the other hand neither argon, helium nor nitrogen at pressures as high as 96 atmospheres had any effect on isolated frog sciatic nerve. However, Carpenter (1954) was able to demonstrate in rats, that six out of seven chemically unreactive gases, which included argon but not nitrogen, diminished excitability, blocked conduction and produced a significant depolarization of isolated peripheral nerve (Table 20). A clear relationship was indicated between narcotic activity and solubility in olive oil and there was an indication of interference with the polarization mechanism or with the movement of ions across cell membranes.

However, as with many such studies the inert gas partial pressures were far in excess of those normally required to produce narcosis. Argon for example induced fibre blockade at 310–340 atmospheres, whereas narcosis is apparent in man at about 3 atmospheres.

Other experiments by Carpenter (1953, 1954, 1955) showed that the relative depressant effect of inert gases to protect 50 per cent of a group of mice from electroshock convulsions (E.D.$_{50}$) was considerably less than that required to block conduction in

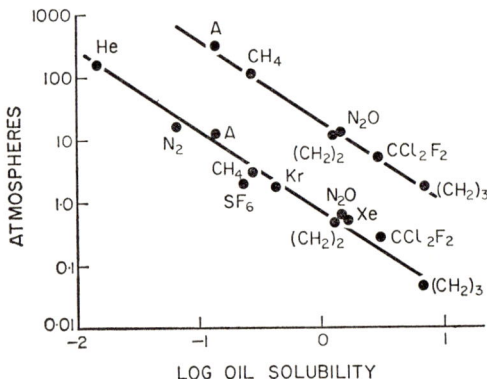

FIG. 6. Relation of E.D.$_{50}$ critical pressures, shown in the bottom line, with the pressures required for reversible nerve fibre blockade in the rat sciatic nerve as shown in the top line (Carpenter 1955).

isolated peripheral nerve (Fig. 6). As the ratio between the two effects is reasonably constant, Carpenter inferred that the mechanism responsible for producing block of conduction in nerve fibres is the same as that in central synapses.

These experiments led to the suggestion that the most likely site of action of inert gases at high pressures is at certain synaptic pathways in the nervous system. That spinal synapses are indeed affected by inert gases has been confirmed by *in vivo* experiments by Chun (1959) and Bennett (1963b). The former, as a result of stimulation of the tibial nerves, obtained electromyograms from the semitendinous muscles of 26 decerebrate cats and studied reflex inhibition in the spinal cord at 7–9 atmospheres of air. The reflex activity of the spinal cord was depressed at these pressures. Depression of the inhibitory process was more rapid than the excitatory process and during decompression the excitatory processes were the first to be restored (Fig. 7).

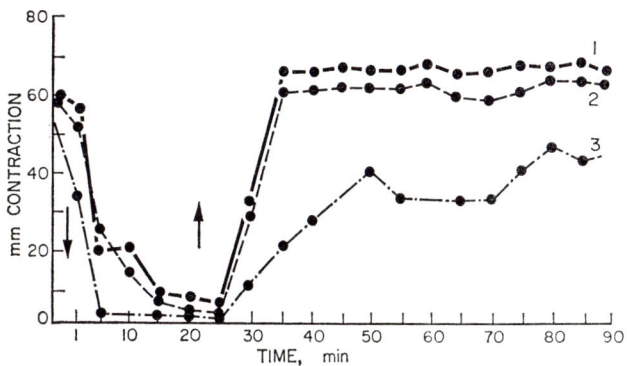

Fig. 7. Progressive change in the inhibitory and excitatory processes of the spinal cord in decerebrate cats at a nitrogen pressure of 9 atm (Chun 1959).
(1) Size of contraction of right muscle (mm).
(2) Size of contraction of left muscle.
(3) Size of the fall of the right muscle on stimulation of the left nerve. The arrows indicate compression and decompression.

The author has recently investigated the effect of increased pressures of argon, nitrogen and helium (220 lb/in^2 inert gas–15 lb/in^2 oxygen, 200 lb/in^2 inert gas–15 lb/in^2 oxygen and 180 lb/in^2 inert gas–15 lb/in^2 oxygen) on spinal synapses and peripheral nerve *in vivo* on 63 rats lightly anaesthetized with Nembutal.

It is not surprising that no change was observed in conduction of sciatic nerve at 220 lb/in² argon (Fig. 8), as the *in vitro* pressures required to effect conduction block were very high (Carpenter 1955). The difference between the pressures is too great to expect any radical effect, even with *in vivo* preparations. Spinal synapses, on the other hand, were readily depressed by inert gases (Fig. 9).

An initial period of depression (B) was followed by augmentation of the potentials (C) after which they were progressively depressed (D, E and F). The extent of the augmentation and depression depended upon the partial pressure of the inert gas and its narcotic potency. Argon mixtures caused the most pronounced effects whereas nitrogen was only mildly narcotic at the pressures used. Helium had no effect. Reference to Fig. 10 shows that the postsynaptic potential was depressed before the presynaptic potential. In fact the latter is augmented after the postsynaptic potential is blocked.

Correlating these results with the observations of Van Harreveld (1944, 1946), Brooks and Eccles (1947), Bishop and McLeod (1954) and Gelfan and Tarlov (1955) the manner in which the spinal synapses are affected by inert gases is very similar to that produced by asphyxia. Brooks and Eccles (1947) concluded that asphyxial blockade in spinal synapses is initially due to failure of propagation in the presynaptic fibres. However, Gelfan and Tarlov (1955) and Lloyd and McIntyre (1949) observed that in asphyxia, reflex failure was the immediate consequence of an asphyxial block of motoneurones. The resistance of the intramedullary afferent fibres to lack of oxygen was found to be considerably greater.

The augmentation in amplitude of the presynaptic intramedullary response (Fig. 10) is not regarded by Gelfan and Tarlov (1955) as a true response. It is thought to be the result of intensification of the current sink responsible for the presynaptic response, which is probably due to the spreading asphyxial block reaching the afferent fibres proximal to the electrodes.

Incomplete ischaemia will cause a block of the motoneurone which suggests that, although oxygen is available, it is not present in the cell in sufficient quantity to meet its metabolic requirements. Such a histotoxic mechanism might well be responsible for inert gas narcosis. The results of investigations of nerve and synaptic transmission do support an asphyxial mechanism. It is known that the cerebral pCO_2 may well be increased at high pressures of inert

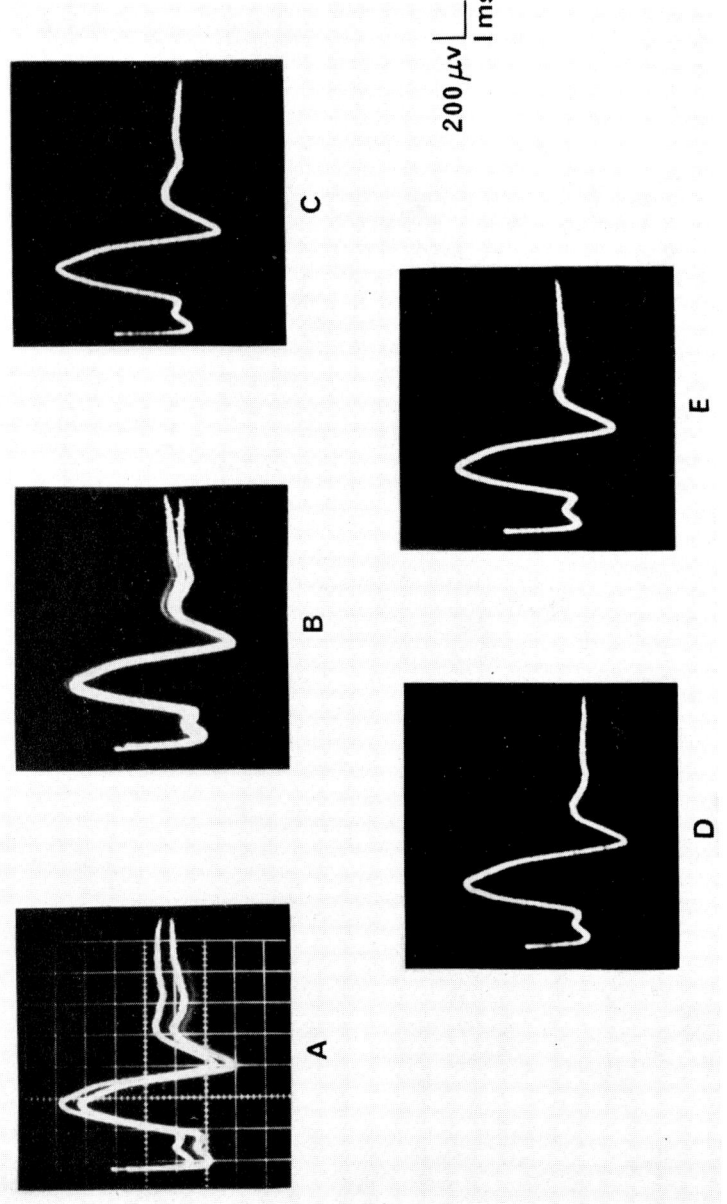

FIG. 8. Rat peripheral nerve conduction at 220 lb A–15 lb O₂: A, At atmospheric pressure. B, After 15 minutes at 220 lb A–15 lb O₂. C, After 30 minutes. D, After 45 minutes. E, After 60 minutes. No effect on conduction time or action potential.

Facing page 46

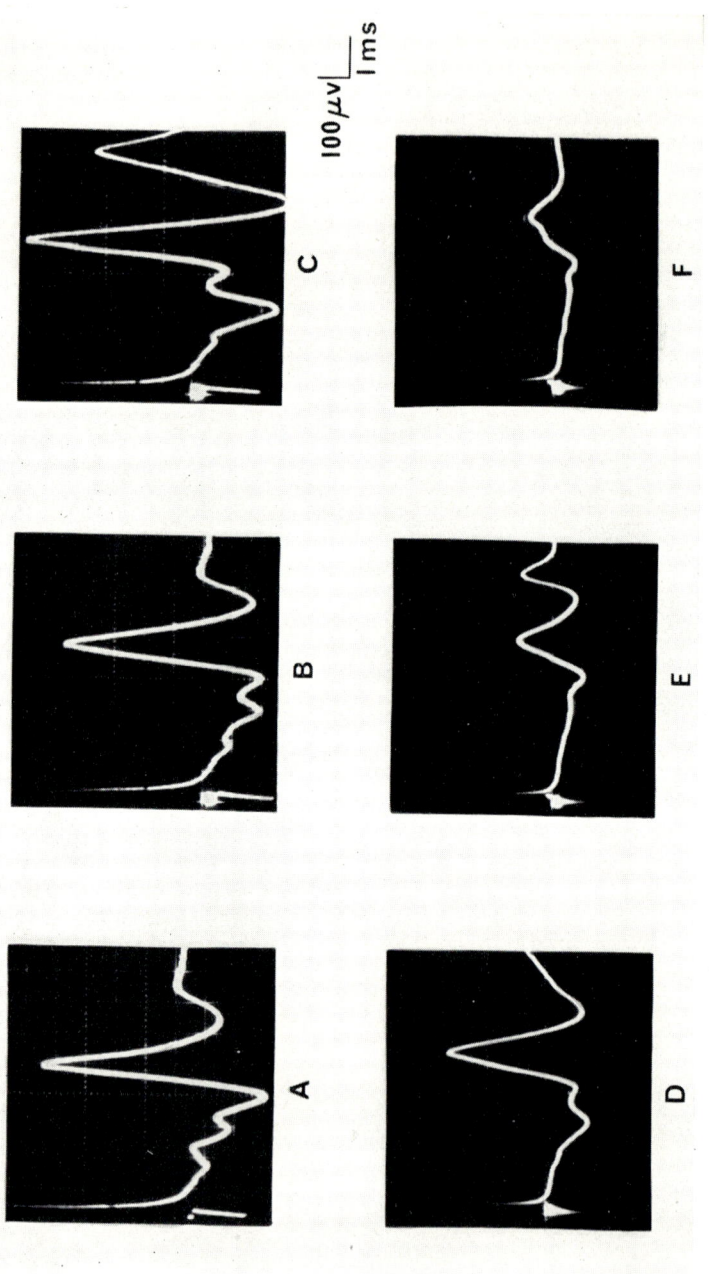

FIG. 9. Rat spinal synaptic conduction at 220 lb A–15 lb O_2: A, At atmospheric pressure. B, After 5 minutes at pressure. C, After 8 minutes, pronounced augmentation. D, After 13 minutes, potentials depressed. E, After 16 minutes. F, After 20 minutes at pressure. Considerable reduction in spike height and increase in conduction time.

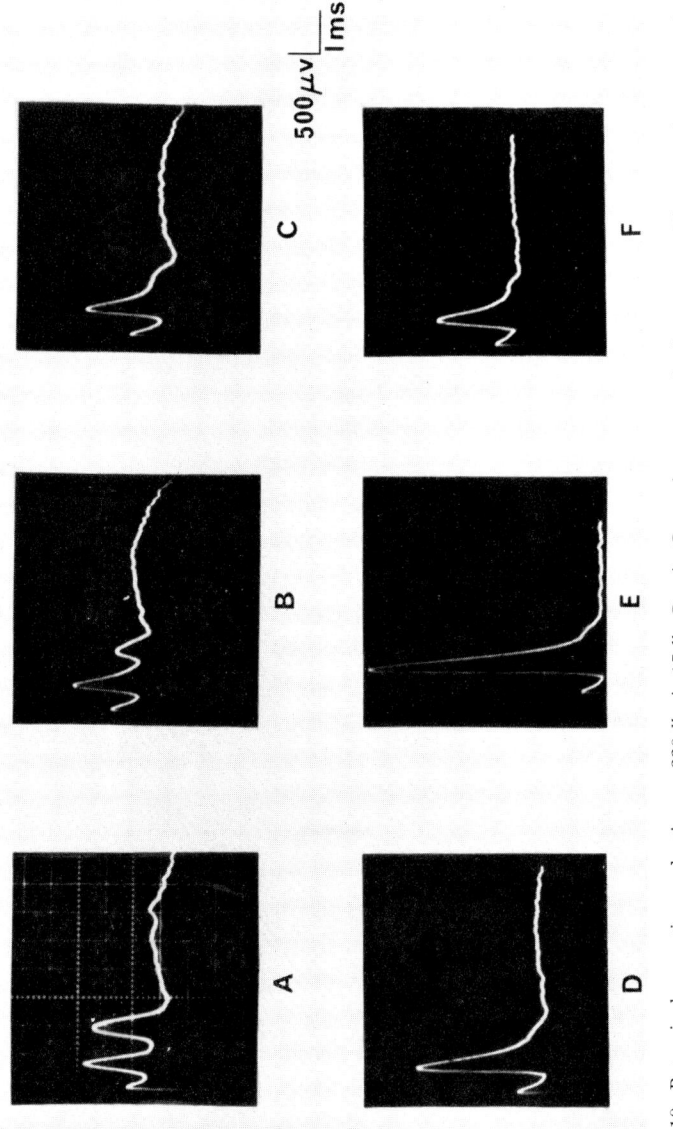

FIG. 10. Rat spinal synaptic conduction at 220 lb A–15 lb O_2: A, Control at atmospheric pressure. First negative potential of presynaptic primary afferent origin. Second negative potential of postsynaptic interneuronal activity. B, After 8 minutes at 220 lb A –15 lb O_2 postsynaptic potential depressed. C, After 15 minutes postsynaptic block. D, Increase of afferent spike due to anoxic block after 18 minutes. E, Further increase of afferent spike at 25 minutes. F, Decompression. Presynaptic potential returned to normal.

gases. The hypoxia however must be subcellular in character as there is no cyanosis with the narcosis.

Although spinal synapses are affected by inert gases it should not automatically be assumed that brain synapses are similarly affected. This is however considered in greater detail elsewhere (Chapter V).

Histotoxic Hypoxia

That a subcellular or histotoxic hypoxia might be concerned with the narcosis is supported by the studies of Ebert, Hornsey and Howard (1958). The sensitivity of bean roots and mouse tumours to ionizing radiation is dependent on the oxygen tension of the tissue. If the oxygen tension is high, then tissue damage results from irradiation. Measurement of the partial pressure of inert gas, added to 1 atmosphere of air required to halve the radiosensitivity of the roots and tumours, led Ebert *et al.* (1958) to conclude that inert gases can displace the oxygen from specific lipid sites probably associated with the cell nucleus or nuclear membrane (Table 21). It is apparent from Table 21 that the

TABLE 21. OXYGEN DEPENDENT SENSITIVITY OF BEAN ROOTS TO IRRADIATION (EBERT *et al.* 1958)

Gas	Pressure in atm for 50% effect	Oil/water solubility
Helium	55	1·7 (37°C)
Hydrogen	55	2·3 (22°C)
Nitrogen	12·5	3·5 (22°C)
Argon	2	4·0 (22°C)
Krypton	2	7·5 (22°C)
Xenon	1·1	14·5 (22°C)

comparative narcotic potency of the various inert gases does not accord with the potencies discussed in Chapter II. However, the general trend is correct and does bear some relation to the oil solubility. Since the mechanism is a cellular one, it may well be unjustified to expect such measurements to agree faithfully with narcotic potencies found for the whole animal.

Read (1958) has suggested that the lipid–water interface, where the displacement adsorption proposed by Ebert *et al.* occurs, is the "perinuclear layer" suggested by Kihlman. This is a lipid–water interface surrounding the nucleus.

Additional support for such a mechanism is provided by Schreiner (1962), who examined the growth rate of the microorganism *Neurospora crassa* in various inert gas environments (Table 22). Adequate oxygen was always present and the carbon dioxide removed. Plotting the growth rate against the square root of the molecular weight gave a straight line relationship.

TABLE 22. EFFECT OF INERT GASES ON GROWTH RATE OF *Neurospora crassa* (SCHREINER 1962)

Inert gas	Mol. wt.	No. of observations	Growth rate (mm/hr)
Helium	4	5	3·51
Neon	20·2	6	3·14
Nitrogen	28	12	2·93
Argon	39·9	10	2·73
Krypton	83·8	11	2·22
Xenon	131·3	12	1·86

Molecular weight was not however considered a pertinent factor, as with sulphur hexafluoride, whose molecular weight is 146·1, the growth rate was 3·03 mm hr. This is within the order for neon and nitrogen with molecular weights of 20·2 and 28 respectively.

As the growth rates were statistically indistinguishable in a high environmental oxygen (Table 23), Schreiner concluded that

TABLE 23. COMPARISON OF HIGH AND LOW OXYGEN PARTIAL PRESSURES ON THE GROWTH RATE OF *N. crassa* IN HELIUM, NITROGEN AND XENON (SCHREINER 1962)

Nominal percentage of gaseous constituents				Number of observations	Average growth rate (mm/hr)
Helium	Nitrogen	Xenon	Oxygen		
95	—	—	5	5	3·51
—	95	—	5	9	2·93
—	—	95	5	12	1·86
50	—	—	50	5	2·98
—	50	—	50	10	3·05
—	—	50	50	10	2·28

the most likely mechanism of the inert gases is to induce a subcellular hypoxia.

Measurements of the cortical available oxygen have been made polarographically (Bennett 1963b, 1965) in an attempt to establish the role of oxygen in the mechanism of narcosis in the whole

MECHANISMS OF THE NARCOSIS   49

animal. The measurements were carried out in conjunction with the pCO_2 determinations and E.E.G. evoked potential studies described in the previous chapter.

It was found that the cortical oxygen availability in cats was affected by the nature of the inert gas breathed (Fig. 11). All the

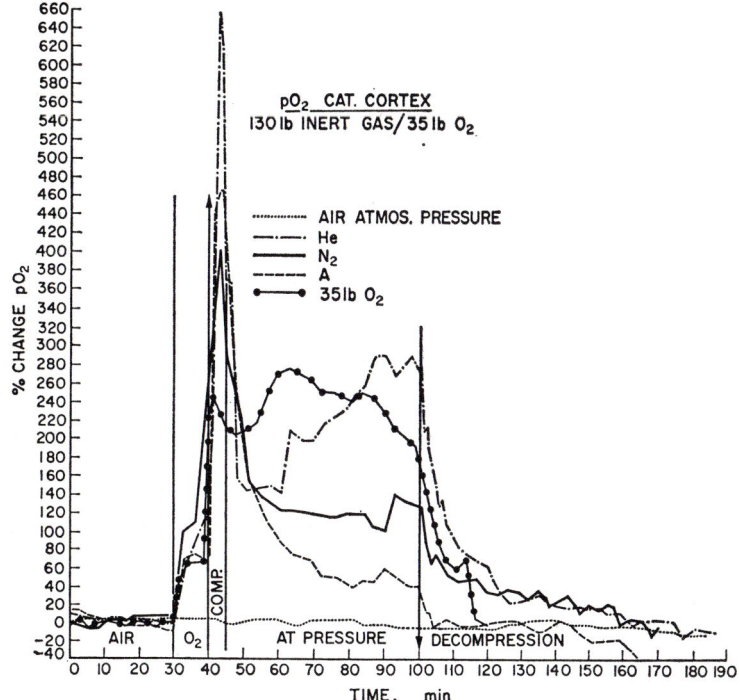

Fig. 11. Cortical available oxygen of chloralosed cats exposed to 130 lb/in² argon, nitrogen and helium in the presence of 35 lb/in² oxygen, 35 lb/in² oxygen alone and at atmospheric pressure. Standard deviations for animals exposed to argon ±36·5 per cent, nitrogen ±23 per cent, helium ±32 per cent, air at atmospheric pressure 1·7 per cent, oxygen alone ±19 per cent (Bennett 1963b, 1965).

tensions were however considerably greater than at atmospheric pressure. The helium–oxygen mixture caused the greatest increase in cerebral oxygen, the nitrogen–oxygen mixture somewhat less and argon–oxygen the lowest. The available oxygen found with the oxygen–helium mixture was in fact little different from that produced by a similar oxygen partial pressure without the presence of any inert gas.

It may be inferred that these results are at least partly the consequence of respiratory embarrassment due to the different densities of the breathing mixtures. Helium appears to have a specific effect of its own. The density of the oxygen–helium mixture is obviously greater than that of oxygen at the same partial pressure alone and yet the cerebral oxygen levels are very similar. The same anomaly with regard to cerebral pCO_2 measurements and the effect of density was considered in the previous chapter. The reasons for these anomalies are at present obscure.

A possible alternative to the density theory is that the differences in oxygen tension are the consequence of differences in oxygen utilization due to a histotoxic hypoxia. A histotoxic hypoxia may be the result of an "active" process of central inhibition as reported by Russek (1962) for sub-toxic doses of sodium cyanide. An increase in the metabolism as a result might increase oxygen utilization. Argon being the most narcotic of the gases examined would utilize the most oxygen and nitrogen rather less. Helium would not affect the tension, so that the cerebral oxygen availability would be similar to that of oxygen alone.

Rather surprisingly, measurements of the cortical available oxygen of cats exposed to mixtures of inert gas and oxygen, where the latter was present only in atmospheric amounts, showed a fall of 20 per cent in brain oxygen, regardless of the inert gas content of the mixture (Fig. 12). Increased metabolism might again be the reason. South and Cook (1953) have reported increases in the metabolism of rat liver, brain and sarcoma slices in the presence of argon, nitrogen and helium. If respiratory embarrassment were a relevant factor, it might have been reasonably assumed that there would have been a correlation between the cerebral oxygen tension and the density of the gas mixture breathed. In fact the helium mixture is more likely to assist respiration than produce respiratory embarrassment (Kane 1940, Dean and Visscher 1941).

If the histotoxic hypoxia theory of narcosis is valid, then it is possible that breathing increased oxygen partial pressures might, in certain circumstances, ameliorate the narcosis, as Schreiner (1962) found with *Neurospora crassa*.

In animals and man such increased oxygen tensions are only a practical proposition with helium and hydrogen. Due to the physical properties of these gases there is not too great a rise in cerebral pCO_2 and quite high oxygen tensions may be obtained without interference by carbon dioxide. With gases such as argon and nitrogen, increasing the pO_2 results also in an increase in pCO_2. Its marked synergistic action may then make the narcosis worse.

Some increase in cerebral oxygen tension might be afforded by assisted ventilation. Buhlman (1961) has maintained that assisted ventilation at increased air pressures will prevent inert gas narcosis.

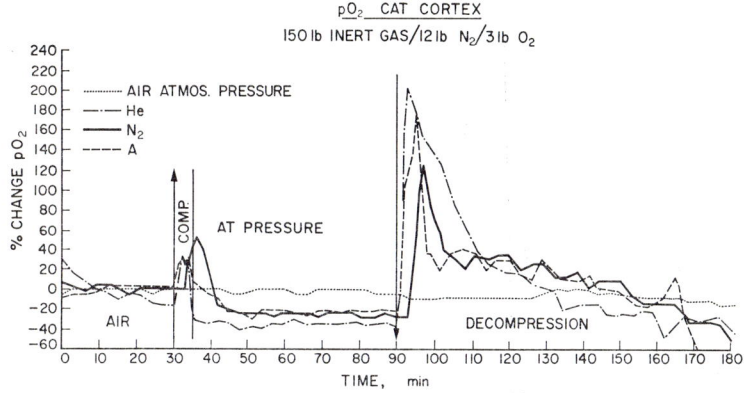

Fig. 12. Cortical available oxygen of chloralosed cats exposed to 150 lb/in² argon, nitrogen and helium in the presence of air and air at atmospheric pressure. Standard deviation for animals exposed to argon ±14 per cent, nitrogen ±6·6 per cent, helium ±14·3 per cent, air at atmospheric pressure ±1·7 per cent (Bennett 1963b, 1965).

The author has examined this possibility by artificially ventilating cats whilst measuring cerebral oxygen availability, carbon dioxide and auditory evoked cortical potentials as the index of narcosis.

Cats exposed to partial pressures of 130 lb/in² argon–85 lb/in² oxygen show a substantial increase in the cortical oxygen tension (Fig. 13) when compared with animals breathing 130 lb/in² argon–35 lb/in² oxygen (Fig. 11). However, the carbon dioxide tension has increased considerably and the evoked potentials are severely depressed.

If such preparations are artificially hyperventilated, by means of a tracheotomy tube and a Palmer pump, the result is very different. As a result of the high ventilation rate, the pCO_2 falls to hypocapneic levels (Fig. 14), and the evoked potentials are not significantly different from controls at atmospheric pressure. The oxygen tension is however considerably reduced, due to the vasoconstriction as a result of the hypocapnia.

Now it is well known that a high pH and low pCO_2, as a consequence of hyperventilation, results in increased cortical excitability as demonstrated by an augmentation of the amplitude of evoked activity (Dusser de Barenne *et al.* 1938, Meyer, Gotoh and Tazaki 1961). Inert gas narcosis is therefore still present and

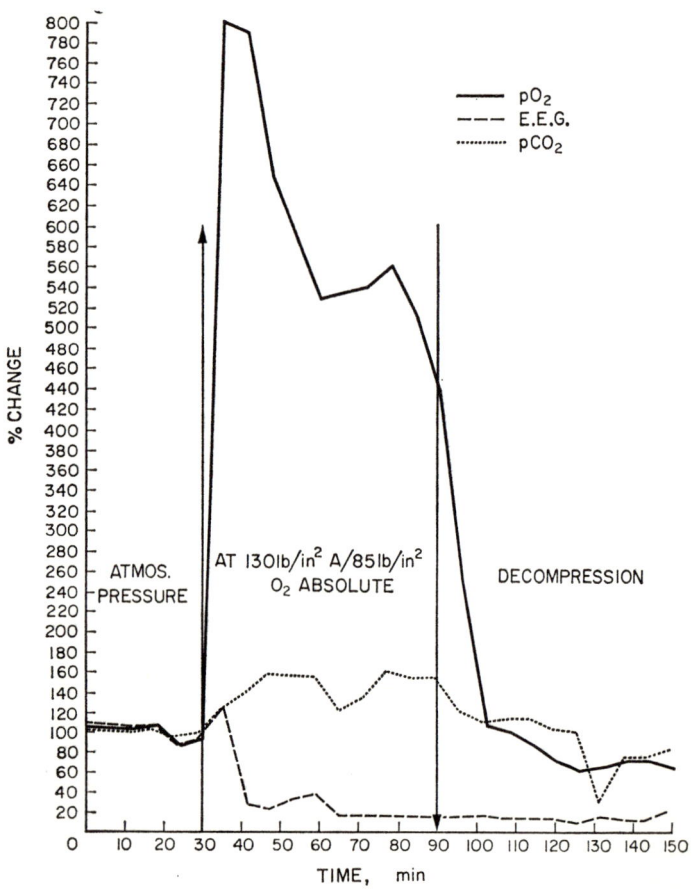

FIG. 13. The effect of a high oxygen partial pressure on the narcotic effect of argon on E.E.G. evoked potential, cortical pCO_2 and pO_2 in the chloralosed cat (45 mg/kg).

the normal evoked potentials are merely the consequence of a reduction in the hypocapneic augmented potential which would normally be found at atmospheric pressure.

Whether or not such changes in acid–base balance would be effective as a means of preventing inert gas narcosis in man remains to be investigated in more detail.

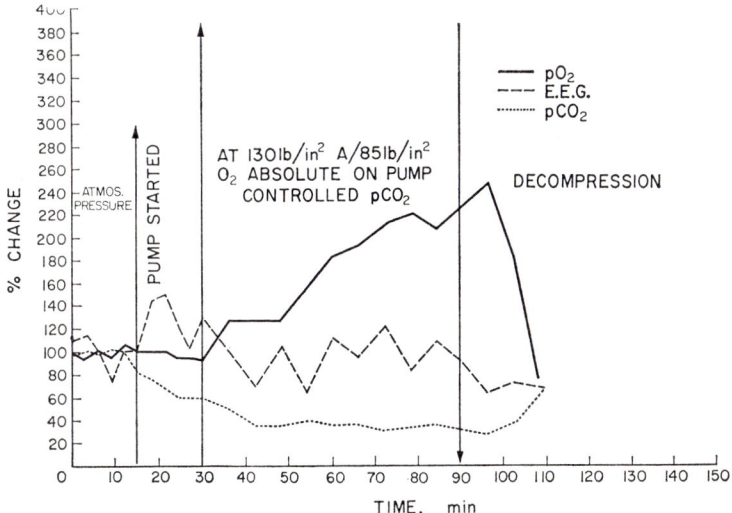

Fig. 14. The effect of hyperventilation on the narcotic action of an argon–oxygen mixture (E.E.G. evoked potential), cortical pCO_2 and pO_2 in the chloralosed cat (45 mg/kg).

If a clear indication of a histotoxic hypoxia is to be obtained from such experiments it is evident that changes in acid–base characteristics due to changes in carbon dioxide tension should not be allowed to interfere. Accordingly the synergistic interference of carbon dioxide was removed by controlling the ventilation rate so that its tension did not differ significantly from that at atmospheric pressure.

Under these conditions it is possible to obtain very high cerebral oxygen tensions without interference by carbon dioxide. The cerebral oxygen tension with the 130 lb/in² argon–85 lb/in² oxygen increases by some 900 per cent but inert gas narcosis is still present (Fig. 15). The slight improvement that may be seen in the depression of the E.E.G. evoked potentials in Fig. 15

compared with Fig. 13 is probably due to removal of the synergistic action of pCO_2 rather than an improvement due to the high oxygen tension. It is evident that oxygen tensions 900–1000 per cent greater than normal are not able to prevent the narcosis. The benefit of assisted ventilation reported by Buhlman may be due to changes in acid–base balance or removal of the synergistic interference of carbon dioxide.

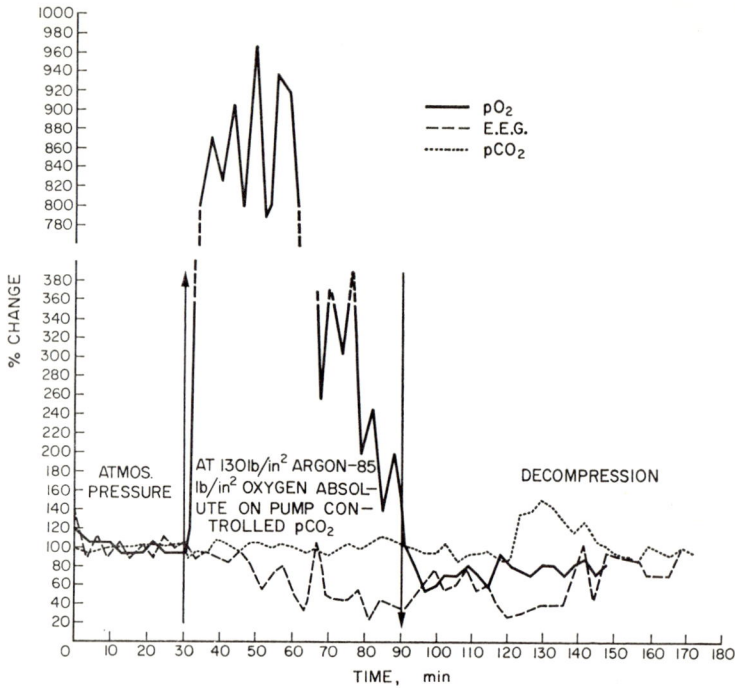

FIG. 15. The effect of a high oxygen partial pressure on the narcotic action of argon on cat cortical evoked potentials (E.E.G.), pCO_2 and pO_2 when the pCO_2 is maintained at control values by means of controlled artificial ventilation.

That the depression of the evoked potentials is indeed a function of the inert gas argon and not some other factor is confirmed by the results of experiments on high oxygen partial pressures alone (Fig. 16). At 85 lb/in² oxygen when the carbon dioxide is controlled by artificial ventilation within normal limits, the auditory evoked potentials of the E.E.G. are not significantly altered although the oxygen tension rises to values in excess of 1000 per cent more than normal.

At lower argon partial pressures, but with the same oxygen partial pressure as the experiments described previously, i.e. 100 lb/in² argon–85 lb/in² O_2, the cerebral oxygen increased by

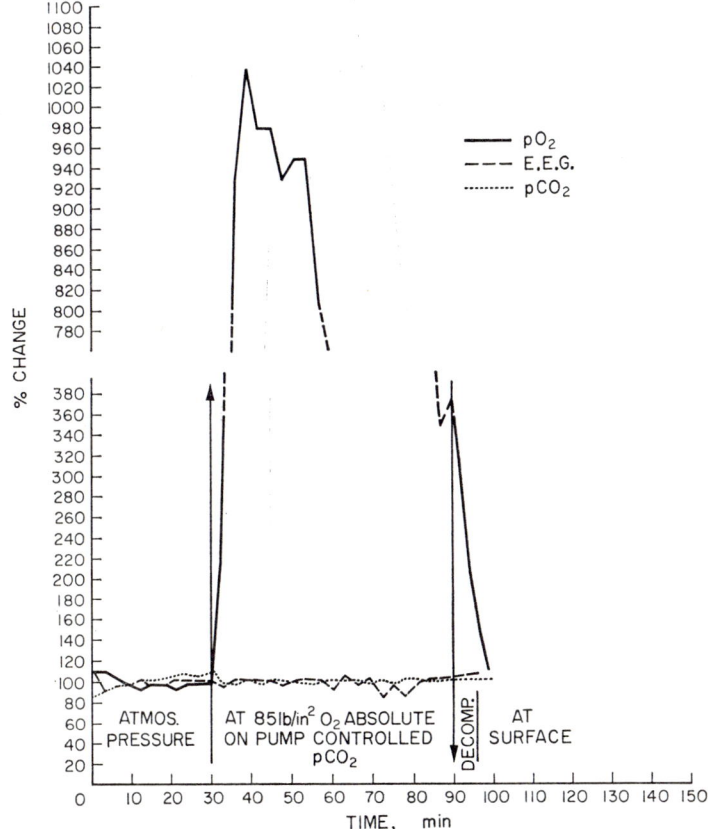

Fig. 16. The effect of a high oxygen partial pressure on the cat cortical pO_2 and evoked potentials when the pCO_2 is prevented from increasing by controlled artificial ventilation.

some 1400 per cent when the carbon dioxide was controlled at normal (Fig. 17). Even this very high oxygen tension, in the presence of a less narcotic concentration of argon, failed to prevent the narcosis as demonstrated by the depression of evoked potentials.

In one case only, when the animal was not ventilated artificially but exposed to similar argon–oxygen partial pressures, the oxygen

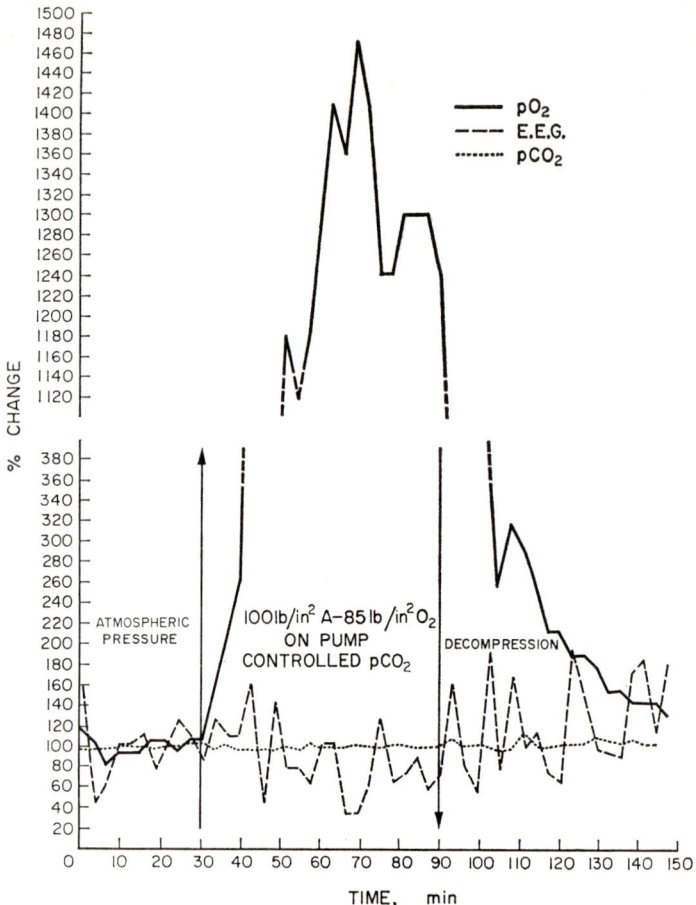

FIG. 17. The effect of a mixture of lower argon partial pressure than in Fig. 15 but with the same oxygen partial pressure on cat cortical evoked potentials and pO_2 when the normal associated rise in pCO_2 is prevented by controlled ventilation.

tension increased to a maximum of 2800 per cent above precompression levels (Fig. 18). This was probably due to the vasodilatation as a consequence of the increased pCO_2. When the oxygen tension exceeded a percentage change of 1300–1400 per cent more than that at atmospheric pressure, the auditory evoked potentials of the E.E.G. returned to control values but were depressed again as the oxygen tension fell.

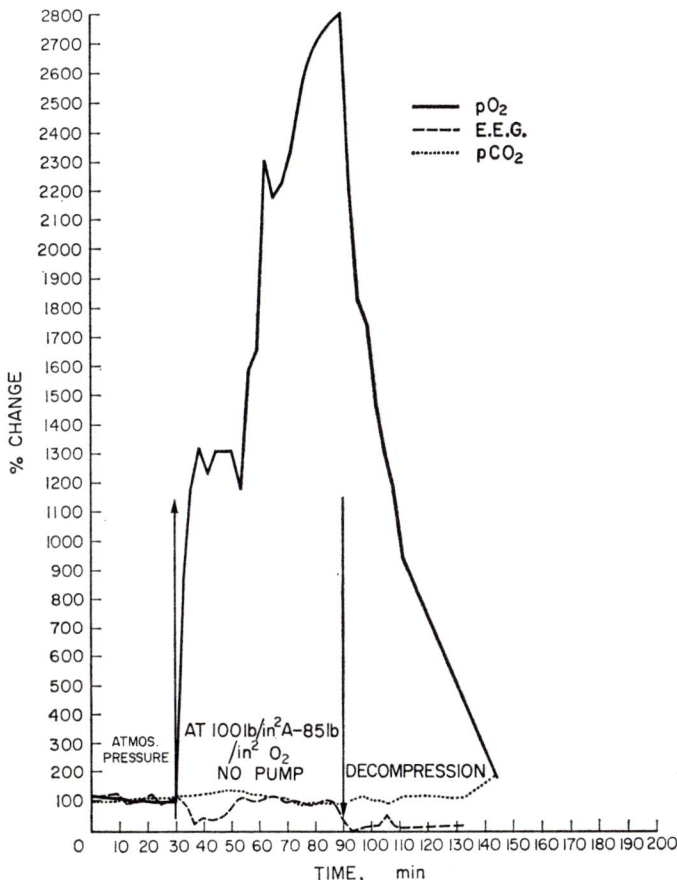

Fig. 18. Chloralosed cat (45 mg/kg) under the same conditions as Fig. 17 but without controlling the ventilation rate.

On the basis of this one result and the experiments described previously it is however not possible to confirm or deny the histotoxic hypoxia theory.

Interference with Metabolism

That asphyxial mechanisms might be the manner by which anaesthetics produce narcosis is not a new conception. Verworn (1903, 1912) and Mansfield (1909) postulated a similar mechanism many years ago. The latter, in particular, related the asphyxial theory to the Meyer–Overton hypothesis of lipid involvement. In connection with the histotoxic hypoxia theory of increased oxygen utilization the Burker (1910) theory should also be considered. Burker on the basis of the results of investigations of the electrolysis of water in the presence of ether concluded that ether utilized oxygen in an oxidative process. No experimental evidence *in vivo* has been obtained for such an inference other than the cerebral oxygen tension measurements described in this chapter and these do not allow a firm conclusion as to the role of oxygen.

Indeed there is considerable *in vitro* evidence (Michaelis and Quastel 1941, Quastel and Wheatley 1932) that there is, on the contrary, a decrease in oxidative metabolism. Studies of the respiration of isolated slices of rat cortex in the presence of glucose showed a marked reduction at concentrations of ether and barbiturates which produce anaesthesia. Under such conditions many anaesthetics inhibit brain oxidation of glucose, lactate and pyruvate, the extent of the inhibition being proportional to the anaesthetic concentration.

It is however questionable whether the depression of oxidation is the cause or the effect of the narcosis or if indeed *in vitro* measurements of minced brain metabolism bear any true relation to *in vivo* mechanisms of anaesthesia (Brazier 1954).

In connection with inert gas narcosis, Levy and Featherstone (1954) examined whether there was a similar *in vitro* metabolism alteration caused by xenon and nitrous oxide on guinea pig brain homogenates and mitochondria. No significant difference in the *in vitro* oxidation of glucose, pyruvate or oxidative phosphorylation was found (Table 24). It was considered as a result that the metabolic theory of narcosis was not applicable to xenon or nitrous oxide.

Similar experiments were carried out by Carpenter (1955, 1956) with much the same result. The utilization rate of the Krebs cycle intermediate, pyruvate, was measured under a variety of

TABLE 24. OXIDATIVE PHOSPHORYLATION IN HOMOGENATES AND MITOCHONDRIA (LEVY AND FEATHERSTONE 1954)

Gas phase	No. of samples	O_2 uptake (microatoms)	Phosphate uptake (microatoms)	P/O
80% N_2–20% O_2 on homogenate	8	2·35	5·3	2·24
Air on homogenate	8	2·05	4·4	2·14
80% Xe–20% O_2 on homogenate	10	2·43	4·8	1·92
Air on homogenate	10	2·66	4·3	1·69
80% N_2–20% O_2 on mitochondria	6	4·8	9·1	1·85
Air on mitochondria	6	4·5	9·1	2·02
80% Xe–20% O_2 on mitochondria	6	4·4	9·0	2·03
Air on mitochondria	6	3·4	7·3	2·14

anaesthetics including xenon, nitrous oxide and cyclopropane in isolated rat sciatic nerve (Table 25). The oxygen consumption of resting nerve was significantly diminished at partial pressures of cyclopropane and nitrous oxide which did not affect electrical excitability. At pressures causing complete blockade of axons,

TABLE 25. PYRUVATE DISAPPEARANCE AND OXYGEN CONSUMPTION OF AXONS AT BLOCKADE CONCENTRATIONS OF ANAESTHETICS (CARPENTER 1956)

	Complete blockade of axons		Pyruvate disappearance		Oxygen consumption	
	Incidence	No. observ.	Inhibition	No. observ.	Inhibition	No. observ.
Phenobarbitol						
5 mM/L.	1	7	30%	7	27%	4
10–15 mM/L.	8	9	62%	7	63%	6
Chloretone						
1·5 mM/L.	0	7	—	—	37%	5
3–5 mM/L.	5	6	30%	4	85%	4
Nitrous oxide						
5 atm.	0	6	—	—	35%	5
10–13 atm.	9	10	5%	5	65%	5
Xenon						
12 atm.	2	2	—	—	57%	3
Cyclopropane						
0·8 atm.	0	4	—	—	24%	8
1·7–2·2 atm.	6	6	5%	5	50%	6

the resting respiration was reduced twice as much. However, the minimum pressure which produced nerve conduction block did not measurably inhibit the disappearance rate of pyruvate.

Carpenter stressed the possibility that the metabolic changes may well be the result rather than the cause of narcosis. As the volume of brain synapses might be estimated as less than 5 per cent, Carpenter considered that inhibition of synaptic metabolism may not significantly affect the overall metabolism of *in vitro* brain preparations. This might well explain why it has proved difficult to obtain experimental support for the metabolic theory of narcosis, for as discussed earlier in this chapter, the synapse is especially sensitive to the depressing effects of inert gases at raised pressures.

In contrast to the evidence of a depressed oxygen consumption, Cook (1950) reported the consumption was accelerated in all stages of the metamorphosis of *Tenebrio molitor* (meal worms) exposed to oxygen–helium and argon–oxygen at atmospheric pressure. Immature forms were more affected than adults and the effect appeared to be connected with nutrition. The oxygen consumption of *Drosophila melanogaster* was depressed by xenon–oxygen mixtures at atmospheric pressure. The results of similar experiments on meal worms by Frankel and Schneiderman (1958) did not however support the observations of Cook.

Nevertheless Cook, South and Young (1951) also reported a significant increase in the oxygen consumption and carbon dioxide output of mice breathing helium–oxygen at atmospheric pressure. Striated muscle, liver, ventricle and sarcoma slices were similarly affected. South and Cook (1953) later confirmed that a 20:80 oxygen–helium mixture increased the rate of oxygen consumption and carbon dioxide production of mouse liver preparations. The respiratory quotient remained unchanged, as the two factors were affected to the same extent.

The oxygen–helium mixture also caused a decrease in the magnitude of inhibition produced by potassium cyanide. South and Cook concluded that helium in some way interferes with the cyanide–cytochrome oxidase bond.

As a result of the further observation that helium did not affect fluoride poisoned liver slices if lactate or pyruvate was the substrate, they further concluded that the glycolytic cycle is the site of both an inhibitory and excitatory effect due to helium.

On the basis of further experiments on the effect of helium on brain and sarcoma slices, Cook and South (1953) concluded that helium indirectly alters the rate of phosphorylation of glucose during anaerobiosis, due to an effect on an unknown factor which varies quantitatively among tissues.

More recently Leon and Cook (1960) have stressed that the increased metabolism of tissues in oxygen–helium mixtures is more likely due to the high thermal conductivity of helium.

Olsen and Klein (1947) have inferred that it is possible for a histotoxic hypoxia to cause changes in the electrical activity of the brain without a change in oxidative phosphorylation but with the additional formation of lactic acid. However, Thomas, Neptune and Sudduth (1963) found that a mixture of nitrogen and oxygen at 5 atmospheres absolute, with the same oxygen partial pressure as at atmospheric pressure, had no significant effect on lactate production (Table 26). The production of lactate was only

TABLE 26. PRODUCTION OF LACTATE AT HIGH AIR PRESSURES (THOMAS, NEPTUNE AND SUDDUTH 1963)

Gas phase	Expt. 1 (μmoles/lactate)	2	3	4
Air (1 atm)	5·3±0·3	4·8±0·3	5·0±0·3	7·6±0·1
N₂ (1 atm)	10·4±0·8	11·4±1·4	8·7±0·2	12·1±0·2
N₂+O₂ (5 atm)				
(a) pO₂ 110 mm Hg	5·6±0·4			
(b) pO₂ 140 mm Hg		4·6±0·3		
(c) pO₂ 36 mm Hg			7·1±0·4	
(d) pO₂ 30 mm Hg				12·1±0·2
No. of experiments	4	4	4	4

increased with nitrogen–oxygen mixtures where the pO_2 was less than 40 mm Hg.

It is clear that the evidence for and against metabolic involvement is conflicting. Indeed, Pittinger and Keasling (1959) rightly claim that no satisfactory theory has yet been proposed relating narcosis to specific biochemical changes.

Oil–Water Phase Reversal

In 1957, Sears and Fenn suggested that the narcosis might be the result of a reversal of oil in water emulsions. Accumulation of inert gases in the lipids of cell membranes in greater concentration than the watery components would render the former more continuous. The membrane would become stabilized and a failure of conduction at the synapse would probably result.

Experiments at 100–130 atmospheres of nitrogen and 60–80 atmospheres of argon did demonstrate such a phase reversal whereas helium was ineffective even at 107 atmospheres (Table 27). Nitrous oxide was anomolous as it would be predicted that it is less narcotic than argon and nitrogen when in fact the opposite is the case.

TABLE 27. EMULSION REVERSAL AT INCREASED PRESSURES (SEARS AND FENN 1957)

Gas	Oil–water solubility 37°C	Oil solubility 37°C	Emulsion reversal pressure (atm)
Argon	5·3	0·14	60
Nitrogen	5·2	0·067	100
Nitrous oxide	3·2	1·6	53
Helium	1·7	0·015	107
Carbon dioxide	1·6	0·876	0·01

Much more evidence is required to substantiate this theory, for again the pressures required to effect the phase reversal are very high compared with the pressures of inert gases that cause narcosis and anaesthesia.

Clathrates

In 1961 Pauling proposed that during narcosis and anaesthesia microcrystals of hydrates of the clathrate type are formed which are stabilized by proteins. In this manner an increase in the impedance of nerve tissue may be caused such that the level of electrical activity of the brain is restricted to that of anaesthesia. At the same time Miller (1961) proposed a similar theory. He suggested that although gas hydrates are not normally formed until excessive pressures are reached an anaesthetic may increase the number of "icebergs" in water. An "iceberg" is defined as an area of highly ordered water surrounding a dissolved gas molecule. Such an ice cover could lower the conductance, stiffen the lipid membrane and occlude the pores of cell membranes.

However, there are many gases which have narcotic or anaesthetic properties which do not form hydrates, e.g. diethyl ether. Miller (1961) in support of his theory quotes helium, hydrogen and neon as gases which are not narcotic and do not form hydrates. In fact both helium and neon will induce narcosis (Marshall 1950).

In conclusion there is insufficient evidence at present available to make any firm decision as to the actual mechanism by which inert gases produce narcosis. However, there is reasonable support for the suggestion that the principal site of action of inert gases is at the synapse and that brain mechanisms are affected by interference with complex polysynaptic connections. How in fact the function of the synapse is affected by inert gases remains a mystery. Several possibilities are discussed in Chapter VI but much more research is required. Although *in vitro* evidence points to a histotoxic hypoxia as the mechanism there is little *in vivo* support and oil phase reversal and clathrate formation are as likely mechanisms.

The extent to which synapses are involved in the brain and the effect of such involvement on the electrical activity of the brain are considered in Chapter V.

CHAPTER V

THE ELECTRICAL ACTIVITY OF THE BRAIN AND INERT GAS NARCOSIS

THAT the narcosis is due to an interference with the function of the brain is obvious. Many workers have therefore investigated the effect of inert gases at pressure on the electrical activity of the brain in the search for a clue to the aetiology of the narcosis. Among the first of these were Rikkl and Krivosheenko (1948), who observed that pressures above 3 atmospheres absolute may cause an inhibition of the function of the cerebral cortex. Inhibition of conditioned reflexes appeared to be a factor.

The Electroencephalogram

Marshall (1951) recorded the electroencephalogram (E.E.G.) of frogs and found that nitrogen at 54 atmospheres reversibly abolished the brain-waves in 15–20 minutes. At 57–61 atmospheres this occurred in only 10–12 minutes, whereas at 13–51 atmospheres there was no significant alteration. This suggested that there was a critical narcotic threshold between 51 and 54 atmospheres.

A similar abolition of brain-wave activity was observed with argon but the pressures required were lower. An argon pressure of 41 atmospheres caused reversible abolition in 20–40 minutes. At the higher pressure of 54 atmospheres, the time to abolition of electrical activity was 13–20 minutes, and at the still higher pressure of 61 atmospheres, abolition occurred in only 11–13 minutes. Pressures from 13 to 37 atmospheres of argon caused no change in brain-wave activity. Similarly, helium at pressures of 41–82 atmospheres exerted no inhibitory effect, even after exposure of the frogs for 8 hours.

These results may be regarded as further supporting the Meyer–Overton theory of narcosis. Although the number of moles of argon dissolved in the lipids is twice that of nitrogen it is within the same order (Chapter 2, Table 12). The critical pressure for nitrous oxide was 2 atmospheres and this gave a C_{lip} (mole/litre) value of 0·117, which was very similar to the value for nitrogen. Again, therefore, there is support for the suggestion that narcosis occurs when a critical concentration of inert gas accumulates at some specific site or sites within the central nervous system. Similar evidence has been found in man by Bennett and Glass (1957, 1961) and is considered in the section concerned with the "inert gas threshold".

Variations in the E.E.G. of cats at a pressure of 5 atmospheres of air (165 feet) whilst under light Dial anaesthesia (25 mg/kg) pointed to neurone hyperexcitability (Jullien, Roger and Chatrian 1953). Neurones were more apt to receive and transmit impulses and the rhythmic repetition of strychnine potentials was facilitated. Experiments with light evoked stimuli indicated that the absolute refractory phase diminished progressively during the first 35–40 minutes of compression and later increased considerably. These changes were however attributed to the raised oxygen pressure and were not related to the narcotic effect of nitrogen.

Roger, Cabarrou and Gastaut (1955) reported similar E.E.G. changes in 12 men exposed to 10 atmospheres absolute of air (300 feet). A diminution in the amplitude of the electrical activity was noted together with an augmentation of the frequency. Evoked potentials, produced by intermittent photic stimulation, were usually of higher amplitude. Unexpected stimuli, normally without effect, were apt to elicit spikes in the region of the vertex. These changes were accompanied by "an enhancement of perceptivity of the global activity of the subject, together with an emotional lability, with a tendency to euphoria".

The enhancement of perception is in contrast to the mass of evidence suggesting that high pressures of inert gases cause a reduction in perception and a generalized depression of brain activity. Yet the results do indeed suggest neurone hyperexcitability. A solution to this anomaly is given in experiments by Morris *et al.* (1955). The latter recorded the E.E.G. of 6 human surgical patients during xenon anaesthesia with additional observations on blood pressure, blood oxygen, blood carbon dioxide and E.E.G. changes.

At first a depression of alpha activity (8–13 c/s) was observed, together with an augmentation of the frequency, followed by a slowing to a rhythmic 5–7 c/s wave-form and some other frequencies (Fig. 19). With deepening anaesthesia, there tended to be an increase in slow activity (2–3 c/s and 4–5 c/s) although many of the higher frequencies remained.

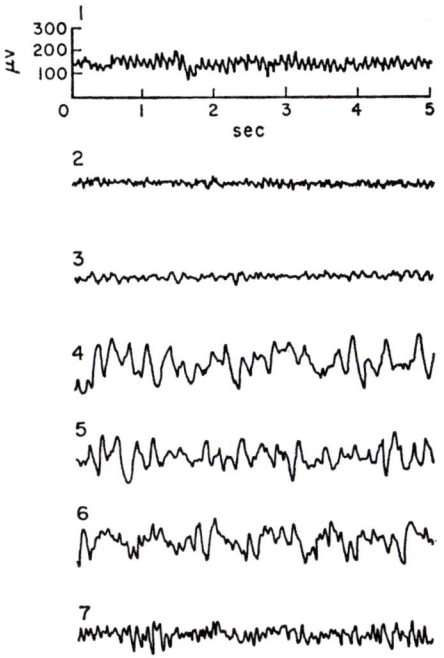

FIG. 19. E.E.G. of patient breathing xenon–oxygen. (1) Breathing air. (2) After 4 minutes breathing xenon–oxygen. (3) After 10½ minutes. (4) After 18 minutes. (5) After 28 minutes. (6) After 68 minutes. (7) Seven minutes after stopping xenon anaesthesia (Morris *et al.* 1955).

The depth of anaesthesia was proportional to the amount of 6 c/s activity. Only three E.E.G. levels were observed rather than the 6 or 7 for cyclopropane and ether anaesthesia. Further, xenon did not appear to produce the very slow rhythmic activity described for these anaesthetics (Courtin, Bickford and Faulconer 1950, Faulconer 1952).

Thus, as with the French workers (Jullien *et al.* 1953, Roger *et al.* 1955) diminution of amplitude and augmentation of frequency were also observed but only during the initial stages of inert gas anaesthesia. Similar E.E.G. changes have been reported in men breathing compressed air at 10 atmospheres absolute (300 feet) by Bennett and Glass (1961) and nitrous oxide at 2 atmospheres absolute (33 feet) by Faulconer, Pender and Bickford (1949). At a pressure of 3 atmospheres absolute (66 feet) nitrous oxide, delta activity occurred.

In agreement with Morris et al. (1955), Pittinger et al. (1955) suggest that the apparent clinical depth of anaesthesia with xenon exceeds the E.E.G. evidence of cortical depression based on scales derived from either ether or cyclopropane.

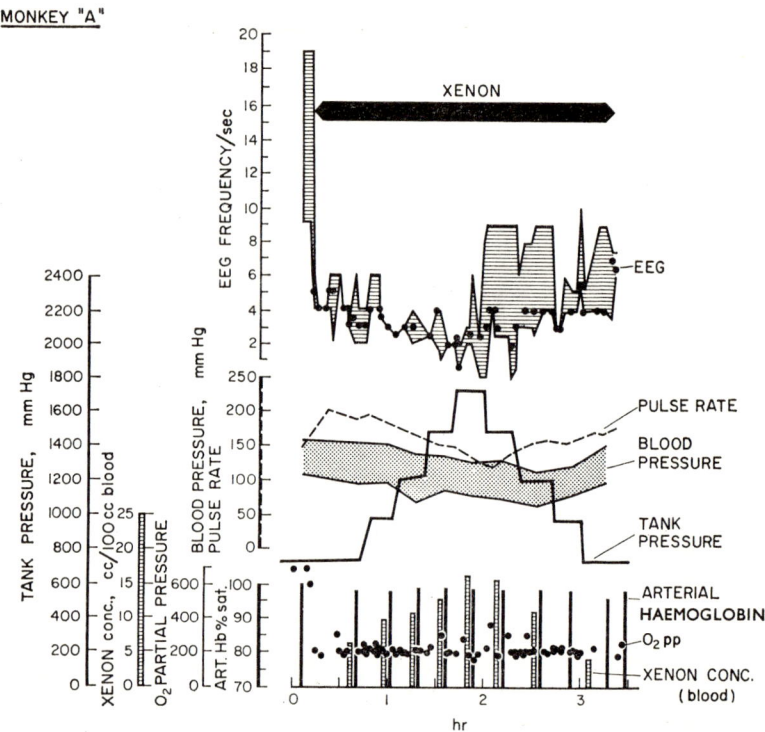

Fig. 20. E.E.G. and other data pertaining to a monkey exposed to xenon anaesthesia at elevated pressures (Pittinger et al. 1955).

Monkeys were exposed to raised pressures of xenon and records of their E.E.G. taken. In addition, pulse rate, blood pressure, arterial blood xenon, respiration rate and arterial haemoglobin were measured (Fig. 20). Profound anaesthesia was observed at raised xenon partial pressures. The E.E.G. showed a decrease in all frequencies. The augmentation of frequency and diminution in amplitude reported for many other inert gases were not observed. Evidently if present, this "excitation" stage is passed through very rapidly during compression.

The results of cuvette oximetry indicated that increasing depth of anaesthesia was not associated with hypoxemia. Similar findings have been made with raised pressures of nitrous oxide by Faulconer *et al.* (1949). Although a fall of up to 8 per cent was observed in the oxygen tension it was not considered significant.

The Italians Albano, Criscuoli and Ciulla (1962a) and Albano and Criscuoli (1962) endeavoured to determine the cause of this controversial hyperexcitability reported by the French workers by exposing six men to various nitrogen–oxygen partial pressures at 10 atmospheres absolute (300 feet). When a mixture was breathed with a high nitrogen partial pressure associated with an oxygen partial pressure equivalent only to that at atmospheric pressure, there occurred signs and symptoms such as a reduction of perception, memory and imagination, decreased vigilance and a tendency to apathy, characteristic of nitrogen narcosis. The E.E.G. showed a decrease in frequency and an increase in voltage with the appearance of theta (4–7 c/s) activity.

If compressed air was breathed at 10 atmospheres absolute, where the oxygen partial pressure was 2·2 atmospheres, theta activity rarely persisted. The narcotic effect of nitrogen was for the most part masked by the hyperexcitability caused by the oxygen. E.E.G. recordings made at 2·2 atmospheres of oxygen alone confirmed that the hyperexcitability was indeed a function of the oxygen partial pressure.

Rate of Compression

A number of workers, including Bean (1950), Frada (1962) and Adolfson (1964), have maintained that if the compression is very rapid, carbon dioxide retention will result. This may cause a decrease in the time of onset of the narcosis and the synergistic effect of carbon dioxide, noted in Chapter III, will increase the severity of the narcosis. Albano (1962) has described 12 cases where the diver had to stop his descent and return to the surface because of severe narcosis at depths which he had attained on former occasions without any trouble. In all these cases the one significant difference was that, where return to the surface was necessary, the rate of descent was especially rapid.

Electroencephalograms of white rats at 21 atmospheres absolute of 96 per cent nitrogen–4 per cent oxygen applied at a rate of 0·6 atm/min have been compared with those of men exposed to 10 atmospheres absolute (300 feet) at a rate of 2 atmospheres/minute (Albano 1962). Whereas nitrogen reached its maximum effect on the brain after 1 hour with the rats, psychometric and E.E.G. changes were found in men after $2\frac{1}{2}$ minutes. The latter time compares favourably with that reported by Wood (1962) as discussed in Chapter II, and Marshall (1951) has reported that mice at 10 atmospheres are narcotized in $1\frac{1}{2}$ hours.

Albano infers that the time of $2\frac{1}{2}$ minutes is probably shorter than that required to reach a concentration of inert gas in the brain sufficient to cause narcosis. This point has been discussed in greater detail in Chapter II in connection with inert gas saturation and it is clear that, on the contrary, much of the brain is over 90 per cent saturated within 2 minutes (Pittinger *et al.* 1954). Further, Albano argues that the contraction of alveolar volume and hypercapnia, with the accompanying desynchronization of the E.E.G. and shortening of the time to the onset of psychometric changes, are proportional to the rate and length of descent. It could however be most misleading to compare the times to the onset of narcosis in rats and man. Such comparisons of different rates should be made in the same animal to be of significant value.

In 114 experiments the response of Wistar rats to the narcotic effect of different nitrogen–oxygen mixtures was examined at three different rates of compression (Fig. 21). The narcotic effect of these mixtures was measured by the increase in voltage required to produce a minimal response, a caudal twitch, to a square wave electric shock (Bennett, Dossett and Kidd 1960). The technique and its evaluation is described fully elsewhere (Bennett 1963b).

It is apparent that the problem of the effect of rate of compression on narcosis is more complex than at first glance. The oxygen partial pressure is of significant importance. Low oxygen partial pressures caused the very slow rate of compression (1 lb/in²/min) to be the most narcotic, whereas at high oxygen partial pressures the fast rate (120 lb/in²/min) was the most narcotic (Fig. 21).

Some confirmation that similar conditions may prevail in man and very fast compression be advantageous rather than the reverse has been obtained in recently completed experiments by Barnard

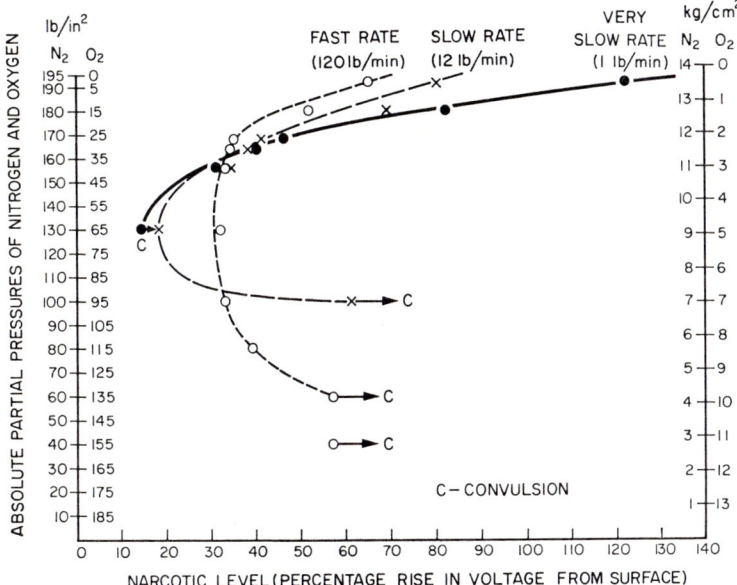

FIG. 21. The mean effects of three rates of compression on the narcotic action in rats of varying nitrogen–oxygen partial pressures (Bennett, Dossett and Kidd 1960).

and the author. Men were compressed on air to 16·1 atmospheres absolute (500 feet) in 20–25 seconds and decompressed 1 minute after leaving atmospheric pressure. Little or no subjective symptoms of narcosis were reported. In fact the general opinion of the men was that a rapid compression was preferable to the normal slow rates of 100 ft/min. There can be little doubt that had the men remained at 500 feet for more than 2 minutes, most would have been unconscious (Wood 1962).

The important factor in these experiments was that the inert gas did not have time to reach the brain in sufficient concentration to cause an appreciable narcosis before decompression commenced. If this rapid compression did cause an accumulation of CO_2 in the tissues, it did not exert any marked narcotic action. However, from the evidence of the experiments in which cerebral pCO_2 was measured at pressure (Chapter III, Figs. 4 and 5) there is reason to believe that cerebral pCO_2 is not increased during compression.

There is however a need for more experiments on the effect of rate of compression on both the narcotic state and alveolar and cerebral pCO_2.

The Nitrogen Threshold

A correlation between changes in the mental state and the E.E.G. and their connection with the ascending reticular formation of the brain-stem has been made by Bennett and Glass (1957, 1961).

At pressure, as the E.E.G. was recorded, subjects were required to solve simple arithmetical problems. Whereas at atmospheric pressure a blocking of the occipital alpha rhythm (8–13 c/s) occurred while the individual was calculating, at increased air pressures no such blocking occurred (Fig. 22). A similar abolition of alpha blocking has been observed in nitrous oxide and oxygen anaesthesia in clinical practice (Clutton-Brock 1961). The time from the start of compression until abolition of the blocking response was inversely proportional to the square of the pressure, i.e. $P \sqrt{T}$ is a constant where $P =$ pressure and $T =$ the time to abolition of blocking (Fig. 23). For example, with one subject, abolition of alpha blocking occurred in 3 minutes at 7 atmospheres absolute (200 feet), 6 minutes at 6·5 atmospheres absolute (150 feet), 12 minutes at 4 atmospheres absolute (100 feet) and 50 minutes at 2·5 atmospheres absolute (50 feet).

Nitrogen was therefore affecting the brain at pressures where subjective narcosis was not apparent. Substitution of an oxygen–helium mixture, after abolition had occurred, resulted in a return of alpha blocking in 4 minutes. Another change noted was a diminution in amplitude of the alpha activity associated with subjective signs and symptoms of narcosis at 200 feet (7 atmospheres absolute). This loss of amplitude was severe at 300 feet (10 atmospheres absolute) and agrees with the first stage E.E.G. changes noted by Morris *et al.* (1955) with light xenon anaesthesia.

The time required at any given pressure to abolish alpha blocking was called by Bennett and Glass "the nitrogen threshold".

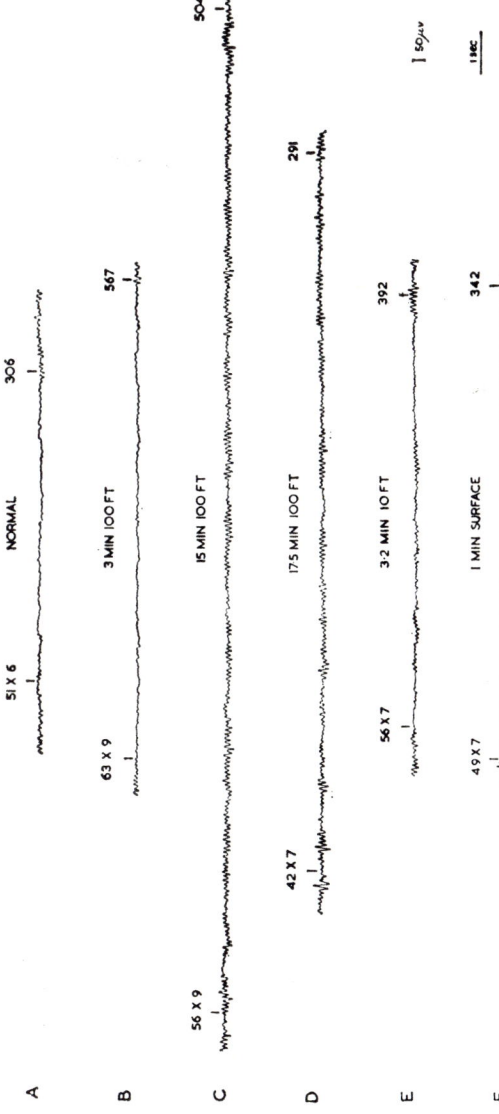

Fig. 22. The effect of 4 atm abs. air on the human E.E.G. whilst the subject answered mental arithmetic problems. Normal alpha blocking response is abolished 15 minutes after the start of compression (Bennett and Glass 1961).

The threshold at 100 feet (4 atmospheres absolute) varied between 5 and 22 minutes depending on the individual. This suggests either a different sensitivity to the effect of nitrogen or a different rate of absorption between individuals.

The results also suggest that even at relatively low pressures, such as 50 feet (2·5 atmospheres absolute), an effect of nitrogen on the brain may be measured. At higher pressures, when subjective appreciation of the narcosis occurs, the effect of nitrogen at specific sites, such as the ascending reticular formation of the brain-stem, may be more severe or the higher tension may be acting more widely as a cortical depressant.

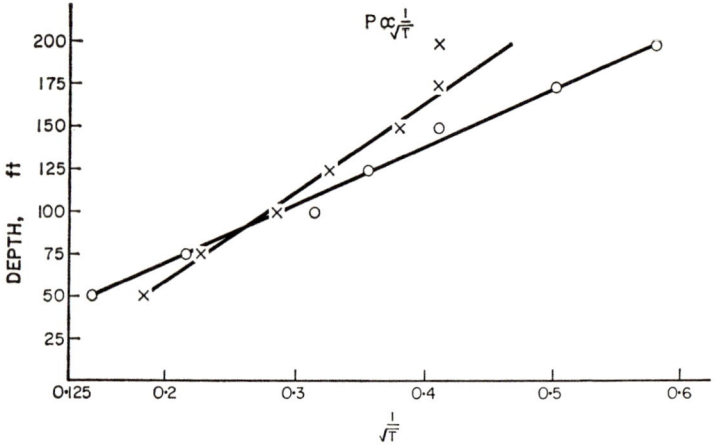

Fig. 23. Relationship between depth of subject and time to abolition of the alpha blocking response for two different subjects (Bennett and Glass 1961).

There was some indication from the results that the alpha blocking abolition was connected with impairment of the mental state as there was a tendency for the answers to the problems to be less correct and take longer to be given after the nitrogen threshold.

The results of a correlation between the degree of performance impairment at psychometric tests with time spent by the individual at a pressure of 4 atmospheres absolute (100 feet) do not however support this tentative indication (Kiessling and Maag 1962). It should be noted that parallel E.E.G. recording was not carried out with the latter experiments. Further, these tests were applied over 3 consecutive 12-minute periods. The fact, therefore, that with 10 men the first period gave an average decrement of 19·3 per cent, compared with 21·2 per cent for the period 12–24 minutes, may not be significant, for the interindividual nitrogen threshold variation is from 5 to 22 minutes. It would thus be possible for the various groups to have the same mean nitrogen threshold. If it was below 12 minutes it would result in all three groups having the same performance decrement.

The E.E.G. changes could only be observed in "R" or responsive type subjects with good alpha blocking at atmospheric pressure. In an attempt to widen the range of subjects, due to the possible practical value of nitrogen threshold determination in selecting divers especially sensitive to nitrogen narcosis or decompression sickness, Bennett (1958) and Bennett and Cross (1960) investigated the use of flicker fusion frequency (F.F.F.). Subjects were asked to assess the critical frequency when a flickering neon light became steady. After a specific time at pressure the frequency at which the flicker became steady changed to a different level, usually one or more cycles per second below the former. As with the E.E.G. method, the time to this change was found to be inversely proportional to the square of the pressure or $P\sqrt{T}=K$.

Experiments were carried out (Bennett 1958) to compare the nitrogen threshold as determined by E.E.G. and F.F.F. in "R" type individuals. The results shown in Table 28 indicate that the

TABLE 28. COMPARISON OF NITROGEN THRESHOLD DETERMINED BY E.E.G. AND F.F.F. (BENNETT 1958)

Subject	Depth in ft	Threshold time to nearest min.	
		F.F.F.	E.E.G.
F	50	38	37
F	150	9	9
C	100	19	19
C	150	15	15
N	100	12	10
N	125	8	8

times to abolition of alpha blocking and a change in the F.F.F. are the same. Support was thus obtained for the theory, first postulated on the basis of the E.E.G. investigations, that a fundamental change occurs in the central nervous system when a certain tension of nitrogen is exceeded.

Lindsley (1958) has suggested that the reticular activating system affects alpha activity, flicker fusion, reaction time and the evoked potentials of the visual areas of cats and monkeys by modifying the excitability cycle of aggregations of cortical neurones. The results of the F.F.F. experiments may therefore be further evidence for involvement of the reticular formation in the mechanism of inert gas narcosis.

It is of interest, in connection with the histotoxic hypoxia theory for the mechanism of the narcosis, that Gellhorn and Hailman (1943, 1944) have also found parallel changes in F.F.F. and E.E.G. during hypoxia. An impairment of alpha blocking was observed during the early stages of hypoxia before, as with nitrogen, any change in E.E.G. frequency. This coincided with a depression of F.F.F.

Other inert gases cause the same changes but the time and depth relationships differ according to the gas (Bennett 1958, 1960). At the same pressure, for example, argon requires only half the time of nitrogen to produce changes in F.F.F. and abolition of alpha activity. It is therefore more correct to call the nitrogen threshold "the inert gas threshold".

Auditory Evoked Potentials

The effect of inert gases on auditory induced evoked potentials has been used by Bennett (1964) to evaluate their comparative effect from electrodes on the cat cortex and in the reticular formation of the brain-stem (Fig. 24). The cats were exposed to a mixture of 130 lb/in^2 inert gas–35 lb/in^2 oxygen for 1 hour. The inert gas content of the mixture was either argon, nitrogen or helium.

Both the argon and nitrogen mixtures depressed the evoked potentials in a characteristic manner (Fig. 25) depending on whether they originated from the cortex or brain-stem. Using the technique of advancing means, graphs of the peak height of the respective components of the evoked potentials were plotted. Differences then became readily apparent.

The positive brain-stem potential was depressed by 65–70 per cent in 25–30 minutes in animals breathing the argon mixture (Fig. 26) and by 60–70 per cent in 15 minutes with the nitrogen (Fig. 27). Helium did not significantly affect the potentials (Fig. 28).

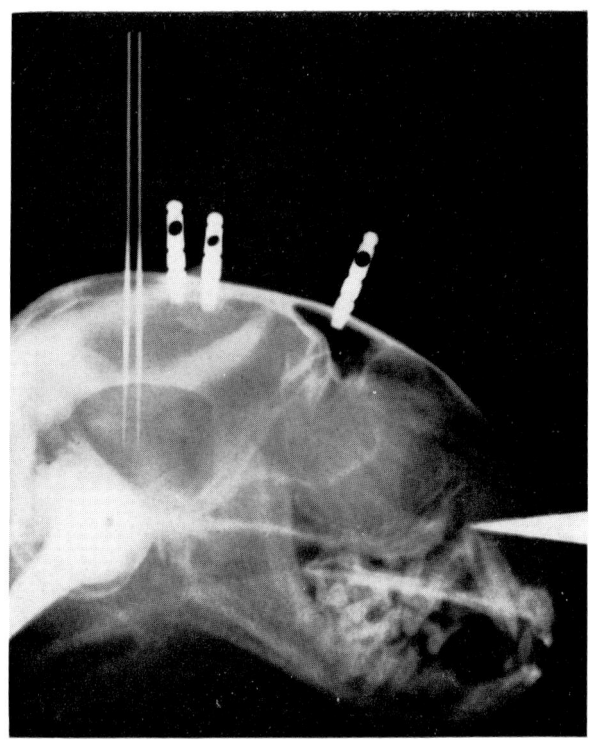

FIG. 24. Lateral X-ray of cat head on stereotaxic instrument with cortical and brain-stem electrodes in position (Bennett 1964).

The positive cortical potential, on the other hand, showed an initial augmentation of 10–25 per cent with argon and nitrogen followed by a depression of only 30–35 per cent in times similar

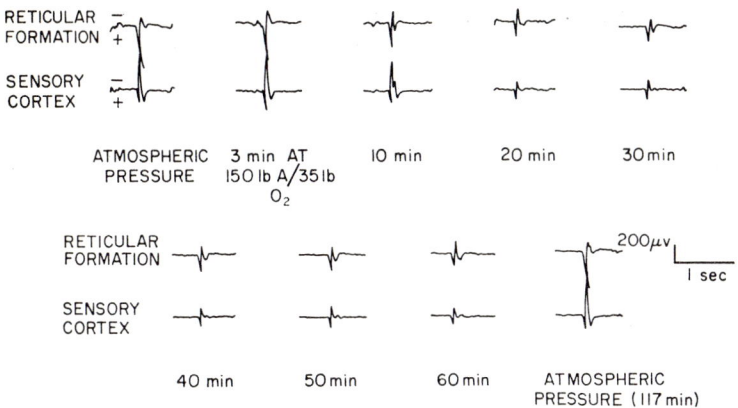

FIG. 25. Evoked potentials at mesencephalic reticular formation and cortex during compression to 150 lb/in² argon–35 lb/in² oxygen absolute. Note the initial excitation followed by depression of the positive cortical potential whilst the cat is at pressure. The brain stem potential shows a depression without the excitation, as does the negative potential of the cortex. The negative potential of the reticular formation is not significantly changed (Bennett 1964).

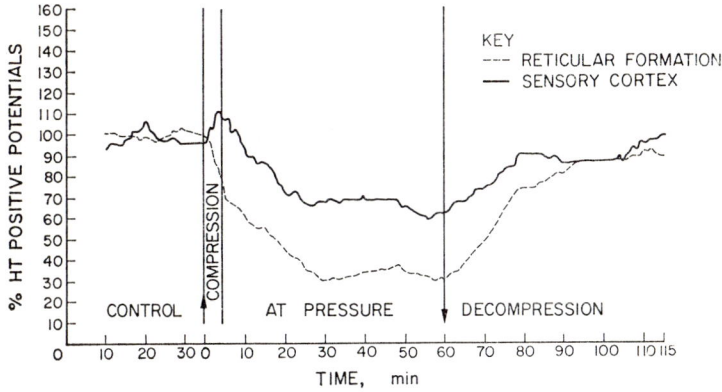

FIG. 26. Advancing means of auditory evoked potentials of cat at 150 lb/in² argon and 35 lb/in² oxygen absolute (Bennett 1964).

to those found in the reticular formation. The negative component of the cortical response was depressed by 80–85 per cent in 25–30 minutes with argon (Fig. 29) and 60–70 per cent in 15 minutes with nitrogen.

It may be concluded that brain synapses are affected by inert gases at increased pressures in a manner similar to spinal synapses

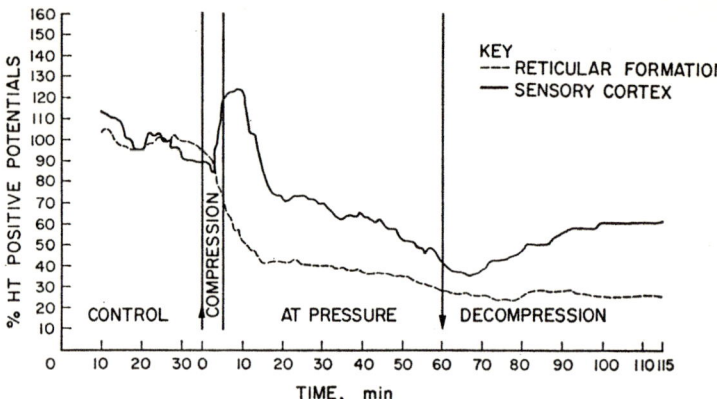

Fig. 27. Advancing means of auditory evoked potentials of cat at 150 lb/in² nitrogen and 35 lb/in² oxygen absolute (Bennett 1964).

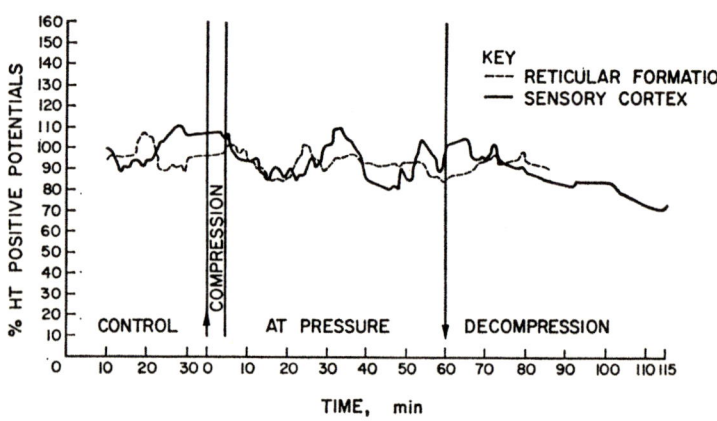

Fig. 28. Advancing means of auditory evoked potentials of cat at 150 lb/in² helium and 35 lb/in² oxygen absolute.

(Chapter IV). Chang (1959) has indicated that the cortical positive potential is of presynaptic and the negative of postsynaptic origin. As the negative potential at the cortex is depressed when the positive potential is augmented, it may be inferred that the depression of brain synapses is initially due to a depression of the postsynaptic processes.

Support was thus obtained for the suggestion that the reticular formation is especially sensitive to raised pressures of the inert gases, due to its large number of synapses. As the cortical negative potential was affected to the same extent, it may be further suggested that the superficial cortical cell bodies and apical dendrites are equally sensitive.

It is apparent that the inert gases, including nitrogen, as with many other agents and anaesthetics such as ether (French, Verzeano and Magoun 1953), cyanide and hypoglycemia (Arduini and Arduini 1954) and nitrous oxide, ethylene and cyclopropane (Davis et al. 1957), act principally on those parts of the nervous system rich in synapses.

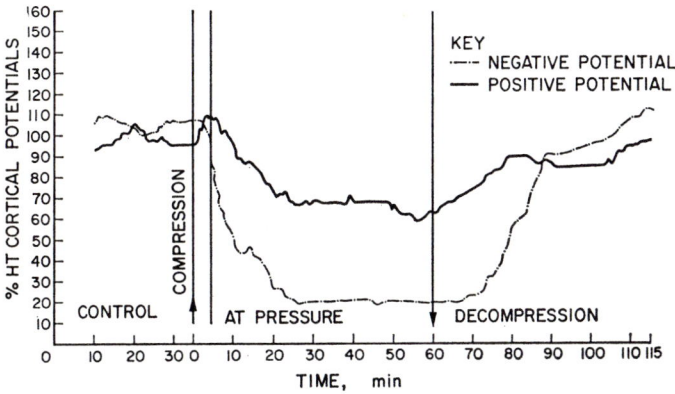

Fig. 29. Advancing means of cortical auditory evoked potentials of cat at 150 lb/in^2 argon and 35 lb/in^2 oxygen absolute (Bennett 1964).

Comparison of the evoked potential studies with similar experiments on the effect of asphyxia and hypoxia on cortical evoked potentials indicates that such mechanisms could well be involved in inert gas narcosis. As there is no hypoxemia the hypoxia would have to be histotoxic. The positive cortical potential of presynaptic origin (Chang 1959) is not as sensitive to oxygen lack as the postsynaptic negative potential (Rice 1960). It may therefore be suggested that the inert gases affect auditory evoked potentials by a mechanism related to the relative oxygen requirements of their generating mechanisms.

Experiments with a mixture of 75 lb/in^2 nitrogen–18 lb/in^2 oxygen (Fig. 30) showed that the auditory evoked potentials at the cortex and brain-stem required about twice the time to reach a maximum level compared with a mixture of 150 lb/in^2 nitrogen–35 lb/in^2 oxygen. This gives added support to the suggestions of Meyer and Hopff (1923), Marshall (1951), Carpenter (1954) and Bennett and Glass (1961) that the narcosis is related to the accumulation of a critical concentration of molecules of inert gas at a specific site in the central nervous system, presumably associated with brain synapses. The possible mechanisms by which this could be effected are considered in the next chapter.

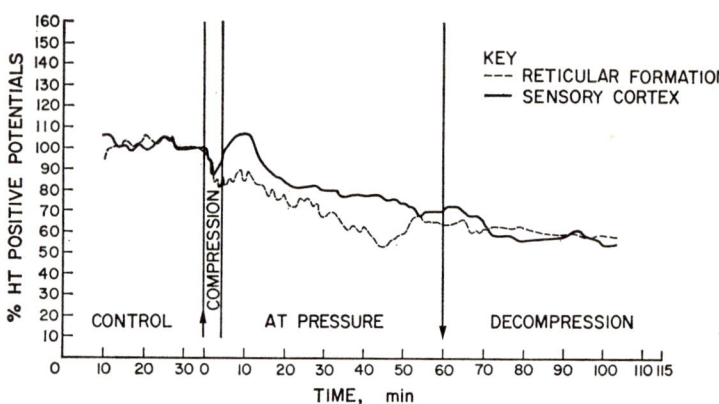

FIG. 30. Advancing means of auditory evoked potentials of cat at 75 lb/in^2 nitrogen and 18 lb/in^2 oxygen absolute (Bennett 1964).

The increased excitability of neurones during the initial moments of narcosis (Jullien, Roger and Chatrian 1953, Roger, Cabarrou and Gastaut 1955) is clearly shown by means of the evoked potential studies. It is not however clear whether this increased excitability is due to a progressive asphyxial depolarization due to a histotoxic hypoxia or to a preferential depression of inhibitory synapses with a consequential release of excitatory mechanisms or to the raised oxygen partial pressure (Albano *et al.* 1962, 1962a).

CHAPTER VI

THE POSSIBLE ACTION OF INERT GASES ON SYNAPTIC MECHANISMS

IT HAS been shown in the previous chapters that inert gas narcosis is probably the result of interference with central synaptic transmission. Possible ways by which this may be achieved will now be considered. Sites of action at the synapse where raised pressures of inert gas may interfere are shown in Fig. 31 at a, b and c.

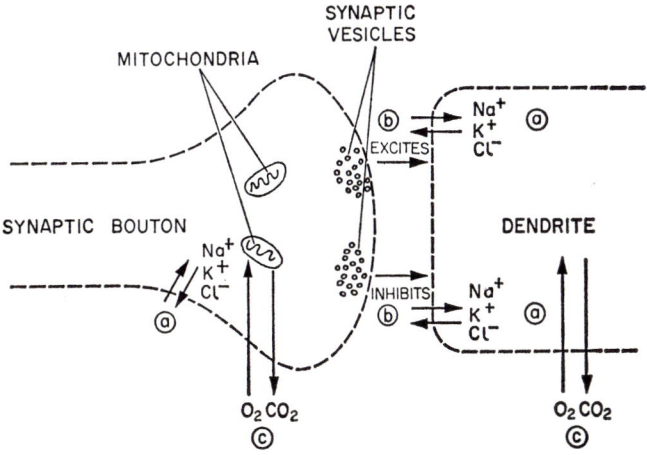

Fig. 31. Possible sites of action of inert gases at the synapse.

If it be supposed that the lipid fraction of the membranes surrounding the presynaptic and postsynaptic cells is packed with inert gas molecules, as is the extracellular fluid and synaptic gap, transmission may be blocked in several ways. The extent and rapidity of onset of the synaptic block would depend upon the lipid solubility, the rate of access and the size of the molecules entering the region of the synapse.

Prevention of the Permeability of Ions through Cell Membranes
(Fig. 31, a)

If inert gas molecules packed into the lipid fraction of the cell membrane, sodium ions would probably find difficulty in penetrating the membrane due to mechanical interference and reduction of pore size (Sears 1962). This is by no means a new theory. It was first suggested by Lillie (1916) and was known as the "permeability theory of anaesthesia".

In 1956 Thesleff found that a number of the non-volatile anaesthetic drugs blocked the production of action potentials in frog skeletal muscle fibres by stopping sodium conductivity across the cell membrane. Yamaguchi (1961) demonstrated an identical suppression of sodium conductivity with chloroform and ether. Thesleff also observed that the dose of the drugs required to produce narcotic effects in whole frogs was similar to that required to affect the action potentials of skeletal muscle fibres. It was therefore suggested that the mechanism of narcosis involved a similar inhibition of sodium conductivity in the neurones of the central nervous system.

Frank and Sanders (1963) examined the effect of procaine, cinchocaine, pentobarbitone, ether and leptazol on the motor activity of intact mice and on the isolated cat cerebral cortex. The first two drugs, when given without the others, caused convulsions and hyperexcitability. However, depression of the central nervous system was caused if the drugs were administered one hour after phenobarbitone. Subanaesthetic doses of the latter abolished the righting reflex of mice previously given convulsant doses of procaine and cinchocaine. Leptazol antagonized this effect. Both procaine and pentobarbitone depressed electrically induced evoked responses in isolated cat cortex, whereas leptazol increased the responses. These and many other experiments led Frank and Sanders (1963) to conclude that "general and local anaesthetics have a fundamentally similar action on neurones in the central nervous system".

However, inhibitory systems are preferentially depressed by inert gases (Chun 1959). The inhibitory postsynaptic potential is considered to be the result of the migration of K^+ and Cl^- ions whose hydrated size is equal to or less than Na^+. The result of the migration of large ions such as Na^+ is facilitatory transmission (Wyke 1960).

If ion permeability is blocked by inert gases it is logical to assume that migration of larger ions would be prevented before smaller ions. Facilitatory mechanisms would then be blocked before inhibitory. As this is not the case, it is difficult to believe that interference with ion permeability is a likely mechanism. It would seem more likely that the failure of ion migration is the consequence, rather than the cause, of the narcosis.

Inhibition of Synaptic Transmitters (Fig. 31, b)

On much the same grounds as those considered in connection with ion permeability, Sears (1962) suggests that a crowding of the lipid molecules of the synaptic membrane would make it difficult for the synaptic vesicles to release their contents.

Such a mechanism would agree with the finding that the postsynaptic neurone is depressed by inert gases before the presynaptic (Chapters IV and V) and Bennett (1962) has reported that intraperitoneal injections of Carbachol will prevent the narcosis in rats.

Again however it is difficult to understand why the vesicles containing inhibitory transmitters should be more easily prevented from releasing their contents than those containing excitatory transmitters. Much still remains to be understood about inhibitory mechanisms. Perhaps it is not the difficulty of release of the transmitters but the difficulty in crossing the synaptic gap with its increased density of inert gas molecules. Alternatively, the narcosis could be the result of interference with the manner in which the synaptic transmitters act on the postsynaptic membrane.

Metabolism and a Histotoxic Hypoxia (Fig. 31, c)

Inert gases at raised pressures could interfere with a number of factors concerned with metabolism. Carpenter (1955) has suggested that 50 per cent of the total phospholipid content of the central nervous system is accounted for by the mitochondria. Perhaps the mitochondria, so responsible for cell metabolism, constitute the lipid phase proposed by Meyer and Overton. A displacement adsorption of oxygen at this site would result in a hypoxia or asphyxial block at synapses. This could readily occur even though the blood was fully saturated with oxygen.

As shown in Chapters IV and V, the nervous system is affected by inert gases in a manner which implies interference with the supply of oxygen needed to maintain its functional integrity. The postsynaptic neurone is the most sensitive to inert gas narcosis and this fact correlates with its preferential sensitivity to oxygen lack. Further, it is well known that inhibitory mechanisms are especially sensitive to oxygen lack as they also are to raised pressures of inert gases.

The synergistic action of carbon dioxide is also explained by a mechanism of a histotoxic hypoxia. A raised carbon dioxide tension will tend to reverse the chemical equilibrium between oxygen and carbon dioxide (Fig. 31, c). This would decrease the entry of oxygen into the cell. Normally the degree of hypoxia caused by this mechanism may not be enough to produce signs and symptoms of narcosis. Combined with a histotoxic hypoxia as a result of the inert gas partial pressure, a synergistic action might well result.

Meyer, Gotoh and Tazaki (1961) have suggested that carbon dioxide narcosis produces slowing of the E.E.G. which correlates with cortical pH rather than the pCO_2. This increased acidity causes a reversible suppression of the "sodium extrusion pump". Hypoxia facilitates the suppression. This also might account for the synergistic action of carbon dioxide. Thus a number of factors may be involved in the mechanism of narcosis. Each of these by itself would be capable of producing narcosis but the end result may well depend upon the effective balance between inhibition of metabolism as a consequence of a histotoxic hypoxia, interference with O_2–CO_2 equilibrium, the prevention of ion permeability and suppression of the sodium pump.

Unsworth (1963) has recently suggested a composite mechanism of narcosis which incorporates the glial-neurone element rather than the synapse. The theory depends on the work of Hyden (1962) which describes the metabolic symbiosis of neurones and their supporting glial cells. Animals subjected to 8 per cent oxygen for 12 hours showed a marked increase in the oxidative enzyme activity of the neurones at the expense of the glial cells, which reverted to anaerobic glycolysis. This caused a drop of efficiency in energy utilization from 55 per cent with the Krebs cycle to 3 per cent with anaerobic glycolysis.

Now the glia contains some 70 per cent lipid and the neurone 5 per cent. The former will therefore adsorb a much higher concentration of inert gas in accordance with the Meyer–Overton theory. Unsworth (1963) suggests that a histotoxic hypoxia, due to displacement adsorption of oxygen by nitrogen, could initiate a similar mechanism. The consequent loss of efficiency would cause a failure of the sodium–potassium pump and a delay in substrate transfer from the capillaries to the neurone through the glial membranes. This would finally lead to a failure of adequate function of the neurone and glia. The brain has one of the highest glial concentrations in the hypothalamus and the reticular formation. That the reticular formation is one of the specific sites of action of inert gases at pressures has been established previously (Chapter V).

It is apparent that the mechanism of inert gas narcosis is complex and may involve one or all of a number of factors. Similarly the cause of the narcosis is also influenced by many factors. This large number of variables is one of the reasons why the aetiology of inert gas narcosis has remained so difficult to determine over the last 100 years. However, with the increased interest in these problems over recent years and the wealth of specialized equipment now available a complete answer should be available within the next decade.

Although the picture as to the cause and mechanism of the narcosis must for the present remain rather confused, there are ways by which the narcosis can be prevented for practical purposes. These are discussed in Chapter VII.

CHAPTER VII

PREVENTION OF THE NARCOSIS

It may be appreciated from the signs and symptoms of depth intoxication discussed in Chapter I that compressed air narcosis can be a considerable hazard to divers as it is also in submarine escape techniques where the submariner is required to breathe compressed air before making his escape. Aqualung diving is becoming increasingly common all over the world as a sport and recreation, in archaeological surveys, salvage techniques and in numerous other ways. The aqualung diver is today diving deeper and deeper due to the development of better equipment and the narcosis problem is likely to become an increasingly frequent hazard.

There are however ways by which the narcosis hazard may be prevented for most practical purposes. The nitrogen of the air may be substituted for a gas of lower narcotic potency such as helium or hydrogen or there are a number of drugs which may help to prevent the condition. Both methods however require to be used with caution and it would be unwise for anybody to try to dive either breathing oxygen–helium or after taking drugs unless they are quite certain of the difficulties such procedures may involve and their remedy.

Helium and Hydrogen

Helium and hydrogen are weak narcotics and for this reason have been used in deep diving procedures (End 1938, Behnke and Yarbrough 1939, Case and Haldane 1941, Zetterstrom 1948, 1949, Bjurstedt and Severin 1948). Although prevention of the narcosis is possible by breathing these gases, during normal diving there are practical disadvantages. Oxygen–hydrogen is explosive and decompression of men breathing oxygen–helium is more hazardous than with air. The high conductivity of heat of these gases necessitates the use of heavy clothing when diving. Storage, voice distortion and the expense of these gases are further considerations which have restricted their more widespread use.

With divers reaching greater depths every year it will also not be long before even diving with oxygen–helium will produce signs and symptoms of narcosis. Marshall (1951) has shown that helium will cause narcosis provided the pressure is sufficient and Keller has reported signs and symptoms of narcosis at 1,000 feet whilst breathing oxygen–helium.

The Effect of Drugs

Few pharmacological investigations have been made and there is a definite need for more study, particularly on the effects of synaptic inhibitory and excitatory drugs and those affecting acid–base balance. Such experiments would add much to our knowledge of the mechanism of narcosis.

Bennett (1963) has investigated the use of Frenquel (α-(4-piperidyl)-benzhydrol hydrochloride, Merrel National Laboratories Ltd.) as a preventative. The drug was produced for use in controlling psychotic hallucinations, confusion, delusions and behaviour disturbances. Due to its lack of side effects, even at high dosage, it seemed especially suitable for the prevention of the less severe signs and symptoms of inert gas narcosis.

The response of 46 Wistar rats to a minimal electric shock, a caudal twitch, was examined in 102 experiments before and after the administration of Frenquel at pressures up to 12·6 atmospheres absolute (380 feet). With an intraperitoneal dose of 40 mg the narcotic effect of a nitrogen partial pressure of 192 lb/in^2 was prevented (Fig. 32), there being no necessity to increase the voltage of a square wave pulse to effect the same caudal twitch as at atmospheric pressure.

The maximum protective action was produced 48 hours after the last 5 mg dose. This action was a true effect of the drug and not the result of acclimatization to the stimulus.

At lower dosage Frenquel was only partially effective. A 20 mg dose caused a reduction of the increase in voltage normally needed to elicit a caudal twitch from 50 per cent to only 20 per cent with a mixture of 180 lb/in^2 nitrogen–15 lb/in^2 oxygen. The

40 mg dose reduced the 50 per cent to only 2–3 per cent, a level not significantly different from that at atmospheric pressure.

Surprisingly, the same 40 mg dose was found as effective in preventing the more severe narcosis produced by argon at a partial pressure of 180 lb/in² (Fig. 33). This supports a biophysical

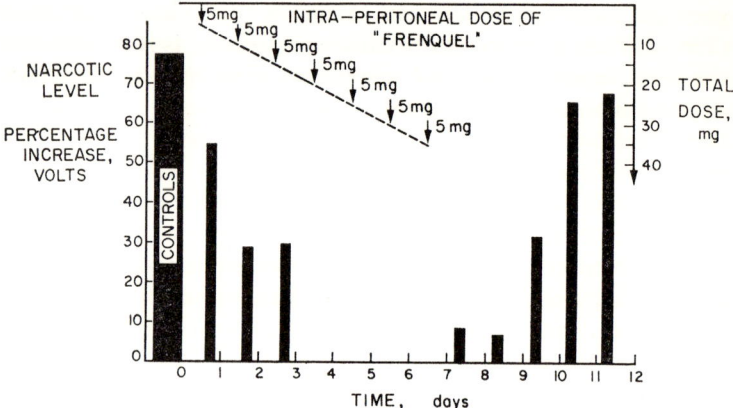

Fig. 32. Effect on the mean narcotic level of successive intraperitoneal doses of Frenquel on 4 rats exposed to a nitrogen partial pressure of 192 lb/in² compared with 15 controls (Bennett 1963).

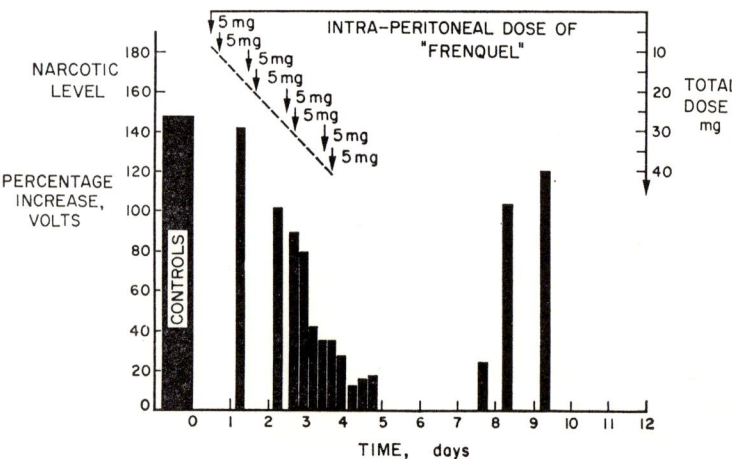

Fig. 33. Effect on the mean narcotic level of successive intraperitoneal doses of Frenquel on 4 rats at an argon partial pressure of 180 lb/in² compared with 7 controls (Bennett 1963).

mechanism for the narcosis, involving perhaps a mechanical barrier to ion permeability or gas exchange across cell membranes. The reason why Frenquel is able to prevent the narcosis is however obscure, for little is known of the mechanism of the drug in clinical use.

Preliminary experiments of the ability of Frenquel to prevent inert gas narcosis in man were carried out on 3 men by Bennett (1961). At 10 atmospheres absolute (300 feet), before and after the oral administration of Frenquel, their performance was examined at arithmetic, letter cancellation and a visual/motor co-ordination test. A psychometric index derived from the answers to the tests established a decrease from 17·5 units at atmospheric pressure to 11·5 at 10 atmospheres absolute (300 feet) of air. After 900 mg/day oral Frenquel, a slight improvement in performance was noted which was not due to acclimatization or learning. The index fell to only 13·5 units and there was some subjective improvement.

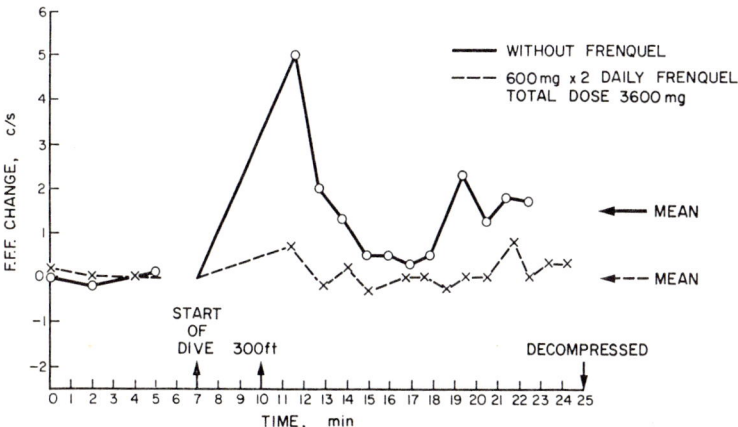

Fig. 34. Critical fusion frequency of flicker at 300 feet (10 atm abs.) before and after administration of Frenquel (Bennett 1961).

The fusion frequency of flicker (F.F.F.) was used to determine the preventive action of a 1200 mg dose. This is the maximum dose used clinically. As may be seen from Fig. 34, Frenquel at this dose prevented the change in F.F.F. normally found on compressing men to such high pressures. Subjectively the narcosis was well controlled.

Although the protective action of the drug is delayed if it is administered orally, experiments showed that intravenous injection afforded a reasonable degree of protection, in rats, within 1 hour.

A series of 11 drugs was investigated for their comparative ability in preventing nitrogen narcosis and oxygen toxicity (Bennett 1962). The electroshock technique, described previously, was used to evaluate the narcotic effect of a nitrogen partial pressure of 180 lb/in^2 on rats and this was compared with the time taken to cause convulsions at 80 lb/in^2 oxygen (Table 29).

Frenquel, Phenacetin, Carbachol, Doriden and Aspirin protected rats against both the effects of high nitrogen partial pressures and the convulsions induced by high oxygen partial pressures. On the other hand, Methedrine, Megimide and Leptazol enhanced the animals' sensitivity to both narcosis and convulsions. Scopolamine induced stupor and some animals died breathing the high oxygen pressure. Physostigmine had no significant effect.

It is possible that the convulsions with oxygen may be due to a "release" mechanism as a consequence of blocking of specific central inhibitory synapses due to a histotoxic hypoxia. Bean (1945) has suggested that the same processes are affected in hypoxia as in oxygen poisoning and that it may be that the toxic action of oxygen is due to a hyperoxic hypoxia. This results in a block of synaptic transmission which preferentially affects inhibitory synapses (Van Harreveld and Marmont 1939).

The mechanisms for oxygen poisoning could therefore be very similar to some of those discussed in previous chapters of this monograph in connection with inert gas narcosis. This could well explain why drugs which control or enhance the effects produced by nitrogen will act in a similar manner with oxygen.

The analgesic nature of a number of the drugs coupled with the electrical stimulus technique requires however that the drugs be examined further by techniques less likely to be affected by the presence of analgesic properties. As with so many of the problems associated with narcosis here is a profitable field of research requiring much more work. The preventive action of drugs on depth intoxication and inert gas narcosis is only in its infancy. Yet here may be the key not only to prevention but also to the mechanisms themselves of this baffling but enthralling problem.

TABLE 29. COMPARATIVE EFFECT OF 11 DRUGS IN PREVENTING OR ENHANCING NITROGEN NARCOSIS AND OXYGEN TOXICITY (BENNETT 1962)

Drug	Controls	Carbachol	Frenquel	Doriden	Phenacetin	Aspirin	Physostigmine	Adrenaline	Scopolamine	Methedrine	Megimide	Leptazol
180 lb/in² N₂-O₂ per cent rise in v	53	0	0	0	7·5	28	52	58	57	61	46	97
	51	0	2	4	5	32	50	42	77	87	66	97
	57	0	5	5	5	19	48	54	57	87	88	75
	57	0	5	5	0	30	51	56	52*	65	92	90
	51	0	2	0	5	33	46	56	57	75	95	74
	49											
15 lb/in²	47											
	52											
	44											
	47											
	48											
	65											
Mean per cent	52	0	2	3	5	28	49	53	60	75	77	87
80 lb/in² O₂ min to convulsion	22	79	33	28	48	31	21	21*	25	18	7	20
	29	55	33	37	52	32	20	13*	37	12	36	6
	24	50	35	45	38	36	12	30	42*	8	6	9
	20	61	33	34	54	28	31	37	24	20	9	6
	23	44	39	32	46	31	21	9	20	18	15	5
	15			34			27					
	24			36			11					
	28			31								
	15			33								
	24			34								
	27			42								
	27			30								
	16											
	23											
Mean minutes	22	58	35	33	48	32	20	22	30	15	15	9

* Animal died.   Means to nearest whole number.

ADDENDUM

CHAPTER VIII

RECENT DEVELOPMENTS

DURING the production of this book, research into the problem of inert gas narcosis has continued and the importance of some of this work in relation to the experiments described in earlier chapters has necessitated the insertion of this addendum.

Onset of Narcosis

Further evidence on the time of onset of narcosis and the effect of rate of compression has been obtained by Bennett *et al.* (1964).

In connection with submarine escape studies, the narcotic effect of very rapid compression with air in only 20 seconds to pressures equivalent to 400–500 feet (13·1–16·1 ats abs.) was studied during a 40-second period at this pressure before decompression to the surface. As the time at pressure was very short, the test of narcosis chosen was two-choice reaction time (Fig. 35).

The procedure required subjects to carry out a control test at atmospheric pressure for 1 minute giving some 25–30 measurements of reaction time per subject. During compression no tests could be made as the subjects were too busy ensuring their eardrums did not rupture. However, during the 40 seconds at pressure, approximately thirteen measurements could be made before decompression began at 5–6 ft/sec. A further 1-minute test was made on return to atmospheric pressure. Additional studies were made to examine any learning error.

At 400 feet a comparison in ten subjects between the expected value of reaction time and that actually obtained showed no significant difference. Ten subjects at 500 feet, however, showed a significant increase in reaction time of 14–15 per cent, accompanied by subjective euphoria (Table 30). One of the subjects, for

Fig. 35. Apparatus for measuring choice reaction time. The experimenter control panel is on the left and that of the subject on the right of the chronotron timer. By changing the switches on the left-hand panel, any of the five different coloured lamps may light up in any of the three different positions.

FIG. 36. Ball-bearing test. A subject has 40 seconds to pick up the ball-bearings one at a time and place them in the tube. The score is the number actually in the tube at the end of that time.

example, reported a temporary hallucination that he was drinking a glass of beer.

Compared with the 20·85 per cent decrement found in two-choice reaction time in ten subjects breathing air at 100 feet (4 ats abs.) by Kiessling and Maag (1962) or the 10 per cent decrement in a single-choice reaction time test in fourteen subjects at

TABLE 30. MEAN RESULTS OF THE EFFECT OF RAPID COMPRESSION TO 400 FEET AND 500 FEET ON THE REACTION TIME

		400 ft	500 ft
1.	Atmos. pressure	40·1 ± 6·8	30·7 ± 4·4
2.	Expected result (learning)	37·8 ± 5·8	29·9 ± 3·7*
3.	Actual result	37·2 ± 4·8	34·2 ± 4·0*
4.	Difference (2−3)	+ 0·06 ± 2·3	+ 4·4 ± 2·2
5.	Return atmos. pressure	36·1 ± 5·8	30·0 ± 3·8

* $P = 0.02$.

150 feet (5·6 ats abs.) found by Shilling and Willgrube (1937), the impairment at 500 feet is not great. That the subjects were experienced instructors from the Escape Training Tank, H.M.S. *Dolphin,* no doubt accounted in part for their low susceptibility. There can be little doubt that novices would have been more severely affected. Here, though, is an example where a very fast compression rate can be advantageous. If the compression had been of the order of 100 ft/min or more time had been spent at pressure, then the narcosis would have been severe (Wood 1962).

The Narcotic Potency of Neon

As described elsewhere (Chapter I), until recently little was known of the narcotic potency of neon. Jean Marshall (1950) appears to be the first worker to have studied its physiological action. In four experiments she found that the threshold for reversible inhibition of reflex activity in the tibial nerve of frogs was between 47 and 61 atmospheres. Thus a mean threshold of 54 atmospheres compares with a threshold of 17 atmospheres for nitrogen and no effect even at 82 atmospheres with helium (Chapter I). More recently Schreiner (1962) has examined the effect of neon on the growth rate of *Neurospora crassa* (Table 22).

Ikels (1964a) has now determined the Bunsen absorption coefficient for neon in water, olive oil and extracted human fat by a gas chromatographic technique. These values have been quoted in Tables 8 and 11. The solubility of neon in water of 0·0094 at 37·6°C is similar to that in the literature (Lawrence *et al.* 1946). The value for olive oil (0·019 at 37·6°C) which has been required for many years in order to predict the narcotic potency of this gas suggests that neon should be less narcotic than nitrogen (0·067 at 37°C) but slightly more narcotic than helium (0·015 at 37°C).

A donation of a quantity of crude neon (80 per cent neon/20 per cent helium) by the British Oxygen Company enabled Bennett (1965) to study the psychometric impairment in ten subjects exposed to 65·6 per cent neon/16·4 per cent helium/18 per cent oxygen at 200 feet (7 ats abs.). A partial pressure of 152 feet equivalent of neon was therefore compared as to narcotic potency with nitrogen at the same partial pressure by exposure to air at 190 feet absolute.

The tests used were a neuromuscular co-ordination test requiring picking up ball-bearings with tweezers and dropping them into a small hole (Fig. 36) and arithmetic. In agreement with its oil solubility, the results confirmed that neon is of low narcotic potency (Table 31). Similar experiments have been carried

TABLE 31. PERCENTAGE CHANGES IN PERFORMANCE IN SUBJECTS EXPOSED TO A NEON PARTIAL PRESSURE OF 4·6 ATS (152 FT) COMPARED WITH A SIMILAR PARTIAL PRESSURE OF NITROGEN

	Nitrogen (Air 190 ft, abs.)	Neon (65·6 Ne/16·4 He/18 O_2 233 ft, abs.)
Sums correct	− 12*	− 3·3†
Sums attempted	− 12*	− 1·7‡
No. of ball-bearings	− 15·6*	+ 2·7§

* P 0·001. † P 0·01. ‡ P 0·05. § Not significant.

out at 300 feet on two subjects when all the tests showed an improved performance compared with air at atmospheric pressure. The mean arithmetic correct was 8 on the surface breathing air and 11·5 at 300 feet breathing crude neon and 20 per cent oxygen. Arithmetic attempted increased from 10 to 12 and the number of ball-bearings from 11·5 to 13. These results suggest that neon is not narcotic at this pressure and that the improvement is the result of learning. Subjectively the divers felt no narcosis. It is interesting to note that whereas the two men breathing the neon mixture had no difficulty during decompression, an attendant breathing air developed a bend and severe itching during decompression.

The poor solubility of neon and its slower diffusion characteristics compared with helium, accompanied by its low narcotic potency, make this a most interesting gas for use in diving or other high-pressure procedures. Research on the practical use of this gas is likely to increase rapidly now that Ikels (1964a) has determined its solubility values. That these values are reliable is emphasized by the work of Dr. E. Smith of the Physical Chemical Laboratory, Oxford, England, who has also measured the oil solubility of neon and obtained a figure of $0·02 \pm 0·002$ (or 21·5 ml neon at NTP in 1000 g olive oil) taking the density of olive oil as 0·92.

Psychometric Impairment Breathing Oxygen Helium

Based on oil solubility calculations, minimal narcosis due to helium would be expected at about 450 feet (14·7 ats abs.) compared with 100 feet (4 ats abs.) breathing air and narcosis equivalent to 300 feet of compressed air should be present at some 1350 feet (41·9 ats abs.). These limits will, however, vary depending on the density and viscosity factors and also on the oxygen partial pressure (see Chapter I).

Baddely (1965) has exposed six subjects to 200 feet breathing 20/80 oxygen/helium in a pressure chamber at the Royal Naval Physiological Laboratory whilst they performed a screwplate test requiring the assembly of nuts and washers together with an arithmetic test. The average decrement in performance was some 10 per cent which was not significantly different from controls at atmospheric pressure. However, in each test one of the six subjects showed an improvement and five did worse.

During the deep-diving studies to 600 feet and 800 feet in men breathing oxygen/helium made at the Royal Naval Physiological Laboratory, Bennett (1965b) has examined the psychometric performance of the men.

Whilst breathing 5/95 oxygen/helium, six subjects were asked to perform the ball-bearing test, multichoice reaction time and arithmetic tests previously described. The tests were carried out after some 15 minutes to compress to 600 feet and then every half-hour subsequently until the start of decompression 4 hours later. A further test was made at 300 feet during the decompression.

Of the six men, four showed a definite performance impairment during the first hour at pressure, which was followed by a slow improvement. This was especially noticeable as regards the ball-bearing test of neuromuscular co-ordination. The psychometric impairment was accompanied by dizziness and in some cases nausea and vomiting during the early stages at pressure. Such symptoms have also been reported at 450 feet and in other dives to 600 feet where the compression was more rapid. In some subjects trembling of the hands and arms and even the whole body is seen.

The two subjects who did not show any marked decrement in performance also reported dizziness but were apparently able to prevent any performance decrement by self-control. The mean results are illustrated in Table 32. Individual results, however, showed a marked impairment with, in some cases, as much as 60 per cent decrement in the ball-bearing test and 50 per cent in arithmetic correct.

That the subjects are worse during the first hour followed by a slow recovery, together with the nausea and tremor, suggests that this may not be classical inert gas narcosis. The oxygen partial pressure is not greater than 1 atm abs. so this would not appear to be a factor. Is this effect due to changes in the acid–base balance of the brain? If so, is the cause a hypocapnia or a hypercapnia? That the divers felt better when working supports the former. Indeed, if the condition were one of classical narcosis, work should have potentiated the narcosis instead of alleviating it.

Additional experiments have been carried out in four subjects breathing 5/95 oxygen/helium at 800 feet (25·2 ats abs.) and a very much more severe decrement in performance was found during the first 20 minutes at this pressure (Table 33). It is obvious that it will not be possible to dive very much deeper under these conditions unless the cause of this impairment can be elucidated.

Among other psychometric studies carried out recently at the

Royal Naval Physiological Laboratory has been an attempt to repeat the findings of Poulton *et al.* (1964) in caisson workers that narcosis can occur at pressures as low as 2 ats abs. (33 feet) (Chapter I). The finding that men exposed to a card sorting test for the first time made more errors than a similar group at atmospheric pressure was not supported by studies of the same test by eighty subjects exposed in groups of ten to air, oxygen/helium and either 40 per cent oxygen/60 per cent nitrogen or 80 per cent oxygen/20 per cent nitrogen at pressures up to 4 ats abs. (100 feet). The experiments were carried out using the "double blind" principle where only very few individuals and none of the experimental operators or the subjects knew what gas was being breathed.

Under such conditions the fact that no narcosis was found suggests that the findings by Poulton *et al.* (1964) of psychological impairment in caisson workers at 33 feet were due to factors other than nitrogen or inert gas narcosis.

TABLE 33. COMPARATIVE PERCENTAGE IMPAIRMENT IN PSYCHOMETRIC PERFORMANCE OF SUBJECTS COMPRESSED TO 600 FEET AND 800 FEET WHILST BREATHING 5/95 OXYGEN/HELIUM

	600 ft (6) (%)	800 ft (4) (%)
Sums correct	− 18	− 42
Sums attempted	− 4	− 6
No. of ball-bearings	− 25	− 53

Among other recent work is the experiment of Fenn (1965) on the effects of oxygen on nitrogen narcosis in *Drosophila* (fruit flies). The fact that at 30 atmospheres of nitrogen the insects were unaffected unless at least 1 atmosphere of oxygen was present suggested to Fenn that this was evidence for an oxygen/nitrogen synergism, i.e. it is a direct effect and not the result of retained carbon dioxide as discussed in earlier chapters of this monograph. His suggestion, as a result of this work, that nitrogen narcosis in man might be diminished by keeping the oxygen tension at its normal atmospheric value is in fact true, but the cause is more likely due to carbon dioxide effects as a result of the increased oxygen tension rather than a direct synergistic effect of oxygen.

H

TABLE 32. Mean percentage change in performance in subjects breathing 95/5 helium/oxygen at 600 feet for 4 hours compared with performance at atmospheric pressure

Test	Surface (air)	600 ft 20 min (%)	600 ft 1½ hr (%)	600 ft 2 hr (%)	600 ft 2½ hr (%)	600 ft 3 hr (%)	600 ft 3½ hr (%)	300 ft (decomp) (%)
Arithmetic (no. correct)	15·67	− 18	+ 1·02	− 9·6	+ 9·6	+ 10·06	+ 7·4	+ 21·25
Arithmetic (no. attempted)	19·17	− 4·2	− 2·61	− 7·0	+ 4·33	+ 6·95	− 0·88	+ 6·6
Ball-bearing (no. of balls)	10·67	− 25	+ 9·37	+ 17·15	+ 26·53	+ 9·37	± 15·5	+ 17·15
Reaction time	1·11	− 0·9	− 12·61	− 18·92	− 15·31	− 12·61	− 15·31	− 20·72

Any further discussion of such mechanisms must, however, await the results of work which is being actively pursued in many countries throughout the world. It is the view of the author, however, that the final answer to the cause and mechanisms of inert gas narcosis is now within our reach.

Addendum References

BADDELY, A. (1965). Personal communication from Applied Psychology Research Unit, Cambridge, England.

BENNETT, P. B. (1965b). Narcosis due to helium, neon and air at pressures between 2 and 25·2 ats abs. (33–800 ft) and the effect of such gases on oxygen toxicity. *Proceedings Symposium Human Performance Capabilities in Undersea Operations,* U.S.N. Mine Defence Laboratory, Panama City, Florida, U.S.A.

BENNETT, P. B., DOSSETT, A. N. and RAY, P. (1964). *Nitrogen Narcosis in Subjects Compressed Very Rapidly with Air to 400 and 500 feet.* Medical Research Council, R.N. Personnel Research Committee Report.

FENN, W. O. (1965). Inert gas narcosis. In Hyperbaric Oxygenation. *Ann. N.Y. Acad. Sci.* **117**, 760.

IKELS, K. G. (1964). *Determination of the Solubility of Nitrogen in Water and Extracted Human Fat.* U.S.A.F. School of Aerospace Medicine, Aerospace Medical Division, Brooks Air Force Base, Texas. Task No. 775801, SAM-TDR-64-1.

IKELS, K. G. (1964a). *Determination of the Solubility of Neon in Water and Extracted Human Fat.* U.S.A.F. School of Aerospace Medicine, Aerospace Medical Division, Brooks Air Force Base, Texas. Task No. 775801, SAM-TDR-64-28.

POULTON, E. C., CATTON, M. J. and CARPENTER, A. (1964). Efficiency at sorting cards in compressed air. *Brit. J. Indust. Med.* **21**, 242.

REFERENCES

Adolfson, J. (1964) Compressed air narcosis. A study of human behaviour at increased ambient pressures. M.Sc. Thesis, The Institute of Psychology, University of Gothenburg, Stockholm.

Albano, G. (1962) Influenza della velocita di discesa sulla latenza dei disturbi neuropsichici da aria compressa nel lavoro subacqueo. Paper read at the 25th National Congress of Medicine, Taormina, 15–18 Oct.

Albano, G. and Ciulla, C. (1962) La sindrome neuropsichica di profondita. Note 1. *Lav. Um.* **14**, 269.

Albano, G. and Criscuoli, P. M. (1962) La sindrome neuropsichica di profondita. Note 4. *Bollettino della Societa italiana di biologia sperimentale* **38**, 754.

Albano, G., Criscuoli, P. M. and Ciulla, C. (1962) La sindrome neuropsichica di profondita. Note 2. *Lav. Um.* **14**, 351.

Albano, G., Criscuoli, P. M. and Ciulla, C. (1962a) La sindrome neuropsichica di profondita. Note 3. *Lav. Um.* **14**, 396.

Arduini, A. and Arduini, M. G. (1954) Effect of drugs and metabolic alterations on brain stem aroused mechanism. *J. Pharmacol.* **110**, 76.

Barnard, E. E. P., Hempleman, H. V. H. and Trotter, C. (1962) Mixture breathing and nitrogen narcosis. *Medical Research Council Report*. R.N. Personnel Research Committee.

Bean, J. W. (1945) Effects of oxygen at increased pressure. *Physiol. Rev.* **25**, 1.

Bean, J. W. (1947) Changes in arterial pH induced by compression and decompression. *Fed. Proc.* **6**, 76.

Bean, J. W. (1950) Tensional changes of alveolar gas in reactions to rapid compression and decompression and question of nitrogen narcosis. *Amer. J. Physiol.* **161**, 417.

Behnke, A. R., Thomson, R. M. and Motley, E. P. (1935) Psychologic effects from breathing air at four atmospheres pressure. *Amer. J. Physiol.* **112**, 554.

Behnke, A. R. and Yarbrough, O. D. (1938) Physiologic studies of helium. *U.S. Nav. Med. Bull.* **36**, 542.

Behnke, A. R. and Yarbrough, O. D. (1939) Respiratory resistance, oil–water solubility and mental effects of argon compared with helium and nitrogen. *Amer. J. Physiol.* **126**, 409.

Bennett, P. B. (1958) Flicker fusion frequency and nitrogen narcosis. A comparison with E.E.G. changes and the narcotic effect of argon mixtures. *Medical Research Council Report*, R.N. Personnel Research Committee.

Bennett, P. B. (1960) Inert gas narcosis. *Ergonomics* **3**, 273.

Bennett, P. B. (1961) A preliminary investigation into the prevention of nitrogen narcosis with Frenquel. *Medical Research Council Report*. R.N. Personnel Research Committee.

Bennett, P. B. (1962) Comparison of the effects of drugs on nitrogen narcosis and oxygen toxicity in rats. *Life Sciences No.* 12, 721.

Bennett, P. B. (1963) Prevention in rats of the narcosis produced by inert gases at high pressures. *Amer. J. Physiol.* **205**, 1013.

Bennett, P. B. (1963b) Neurophysiologic and neuropharmacologic changes in inert gas narcosis. *Proceedings 2nd Symposium on Underwater Physiology*. National Academy of Sciences, Washington, Publication 1181, p. 209.

REFERENCES

BENNETT, P. B. (1964) The effects of high pressures of inert gases on auditory evoked potentials in cat cortex and reticular formation. *Electroenceph. clin. Neurophysiol.* **17**, 388.

BENNETT, P. B. (1965) Cortical CO_2 and O_2 at high pressures of argon, nitrogen, helium and oxygen. *J. Appl. Physiol.* **20**, No. 6.

BENNETT, P. B. and CROSS, A. V. C. (1960) Alterations in the fusion frequency of flicker correlated with electro-encephalogram changes at increased partial pressures of nitrogen. *J. Physiol.* **151**, 28–29 P.

BENNETT, P. B., DOSSETT, A. N. and KIDD, D. J. (1960) Effect of rate of increasing pressure on the narcotic effect of oxygen and nitrogen. *Medical Research Council Report.* R.N. Personnel Research Committee.

BENNETT, P. B. and GLASS, A. (1957) High partial pressures of nitrogen and abolition of blocking of the occipital alpha rhythm. *J. Physiol.* **138**, 18–19 P.

BENNETT, P. B. and GLASS, A. (1961) Electroencephalographic and other changes induced by high partial pressures of nitrogen. *Electroenceph. clin. Neurophysiol.* **13**, 91.

BERT, P. (1878) *La Pression Barométrique.* College Book Co., Columbus, Ohio, 1943.

BISHOP, P. O. and McLEOD, J. G. (1954) Nature of potential associated with synaptic transmission in lateral geniculate body of cat. *J. Neurophysiol.* **17**, 387.

BJURSTEDT, H. and SEVERIN, G. (1948) The prevention of decompression sickness and nitrogen narcosis by the use of hydrogen as a substitute for nitrogen. *Milit. Surg.* **103**, 107.

BRAZIER, M. A. B. (1954) The action of anaesthetics on the nervous system. In *Brain Mechanisms and Consciousness* (Edited by ADRIAN, BREMER and JASPER), Blackwell, p. 163.

BRINK, F. and POSTERNAK, J. M. (1948) Thermodynamic analysis of the relative effectiveness of narcotics. *J. Cell. Comp. Physiol.* **32**, 211.

BROOKS, C. and ECCLES, J. C. (1947) Electrical investigations of the monosynaptic pathway through the spinal cord. *J. Neurophysiol.* **10**, 251.

BUHLMAN, A. A. (1961) The respiratory physiology of deep sea diving. *Schweiz. Med. Wschr.* **91**, 774.

BUHLMAN, A. A. (1963) Deep diving. *The Undersea Challenge.* The British Sub Aqua Club, London, p. 52.

BURKER, K. (1910) Eine neue Theorie der Narkose. *München med. Wchnschr.* No. 27.

BURNETT, W. A. (1955) The problem of nitrogen narcosis. *J. Roy. Nav. Med. Serv. London* **41**, 188.

BUROS, O. K. (1949) *The Third Mental Measurements Year Book,* Gryphon Press, p. 656.

BUTLER, T. C. (1950) Theories of general anaesthesia. *Pharmacol. Rev.* **2**, 121.

CABARROU, P. (1959) L'ivresse des grandes profondeurs lors de la plongée à l'air. Report for Group d'Etudes Recherches Sous-Marine, Toulon.

CABARROU, P. (1964) L'ivresse des grandes profondeurs. *La Press Médicale* **72**, 793.

CARPENTER, F. G. (1953) Depressant action of inert gases on the central nervous system in mice. *Amer. J. Physiol.* **172**, 471.

CARPENTER, F. G. (1954) Anaesthetic action of inert and unreactive gases on intact animals and isolated tissues. *Amer. J. Physiol.* **178**, 505.

CARPENTER, F. G. (1955) Inert gas narcosis. *Proceedings 1st Underwater Physiology Symposium* (Edited by LOYAL G. GOFF), Public. 377 National Academy of Sciences, Washington, p. 124.

CARPENTER, F. G. (1956) Alteration in mammalian nerve metabolism by soluble and gaseous anesthetics. *Amer. J. Physiol.* **187**, 573.

CASE, E. M. and HALDANE, J. B. S. (1941) Human physiology under high pressure. *J. Hyg. (Cambridge)* **41**, 225.

CHANG, H. (1959) The evoked potentials. *Handbook of Neurophysiology* (Edited by JOHN FIELD), vol. 1, p. 299.

REFERENCES

Chun, C. (1959) Effect of increased nitrogen pressure on spinal reflex activity. *Fiziol. Zhur. S.S.S.R.* **45**, 605.

Clutton-Brock, J. (1961) Electroencephalography. *Brit. J. Anaesth.* **33**, 205.

Cook, S. F. (1950) Effect of helium and argon on metabolism and development. *J. Cell. Comp. Physiol.* **36**, 115.

Cook, S. F. and South, F. E. (1953) Helium and comparative *in vitro* metabolism of mouse tissue slices. *Amer. J. Physiol.* **173**, 542.

Cook, S. F., South, F. E. and Young, D. R. (1951) Effect of helium on gas exchange of mice. *Amer. J. Physiol.* **164**, 248.

Courtin, R. F., Bickford, R. G. and Faulconer, A. (1950). Classification and significance of electro-encephalographic patterns produced by nitrous oxide–ether anesthesia during surgical operations. *Proc. Staff Meet. Mayo Clin.* **25**, 197.

Cousteau, J. Y. (1953) *The Silent World*, The Reprint Society, London.

Cullen, S. C. and Gross, E. G. (1951) The anaesthetic properties of xenon in animals and human beings with additional observations on krypton. *Science* **113**, 580.

Cullen, S. C. and Pittinger, C. B. (1952) Clinical and laboratory observations of the use of xenon for anaesthesia. *Surgical Forum*, p. 361.

Damant, G. C. C. (1930) Physiological effects of work in compressed air. *Nature* **126**, 606.

Danielli, J. F. (1950) *Cell Physiology and Pharmacology*. Elsevier.

Davis, H. S., Collins, W. F., Randt, C. T. and Dillon, W. F. (1957) Effect of anaesthetic agents on evoked central nervous system responses: gaseous agents. *Anesthesiology* **18**, 634.

Dean, R. B. and Visscher, M. B. (1941) The kinetics of lung ventilation. An evaluation of the viscous and elastic resistance to lung ventilation with particular reference to turbulence and the therapeutic use of helium. *Amer. J. Physiol.* **134**, 450.

Dunn, J. A. (1962) Psychomotor functioning while breathing varying partial pressures of oxygen–nitrogen. Report 62-82, School of Aerospace Medicine, Brooks Air Force Base, Texas.

Dusser de Barenne, J. G., Marshall, C. S., McCulloch, W. S. and Nims, L. F. (1938) Observations on the pH of the arterial blood, the pH and the electrical activity of the cerebral cortex. *Amer. J. Physiol.* **124**, 631.

Ebert, M., Hornsey, S. and Howard, A. (1958) Effect on radiosensitivity of inert gases. *Nature* **181**, 613.

End, E. (1938) The use of new equipment and helium gas in a world record dive. *J. Ind. Hyg. Tox.* **20**, 511.

Faulconer, A. (1952) Correlation of concentrations of ether in arterial blood with electroencephalographic patterns occurring during ether–oxygen and during nitrous oxide, oxygen and ether anaesthesia of human surgical patients. *Anesthesiology* **13**, 361.

Faulconer, A., Pender, J. W. and Bickford, R. G. (1949) The influence of partial pressure of nitrous oxide on the depth of anesthesia and the electroencephalogram in man. *Anesthesiology* **10**, 601.

Featherstone, R. M. (1960) Anesthesia: Xenon. *Med. Phys.* (Edited by Otto Glasser), Yearbook Publishers, Chicago, vol. 3, p. 22.

Featherstone, R. M. and Muehlbaecher, C. (1963) The current role of inert gases in the search for anaesthesia mechanisms. *Pharm. Rev.* **15**, 97.

Ferguson, J. (1939) The use of chemical potentials as indices of toxicity. *Proc. Roy. Soc.* B **197**, 387.

Ferris, E. B., Molle, W. E. and Ryder, H. W. (1942) Nitrogen exchange in tissue components of man. Committee on Aviation Medicine of the O.S.R.D. National Research Council, U.S.A. Report No. 60.

REFERENCES

FRADA, G. (1962) Address to the International Symposium on Underwater Medicine Ustica (14–16 Sept.) Folio Med.

FRANK, G. B. and SANDERS, H. D. (1963) A proposed common mechanism of action for general and local anaesthetics in the central nervous system. *Brit. J. Pharmacol.* **21,** 1.

FRANKEL, J. and SCHNEIDERMAN, H. A. (1958) The effects of nitrogen, helium, argon and sulphur hexafluoride on the development of insects. *J. Cell Comp. Physiol.* **52,** 431.

FRANKENHAEUSER, M., GRAFF-LONNEVIG, V., and HESSER, C. M. (1960) Psychomotor performance in man as affected by high oxygen pressure (3 atm). *Acta physiol. scand.* **50,** 1.

FRANKENHAEUSER, M., GRAFF-LONNEVIG, V., and HESSER, C. M. (1963) Effects on psychomotor functions of different nitrogen–oxygen gas mixtures at increased ambient pressures. *Acta physiol. scand.* **59,** 400.

FRENCH, J. D., VERZEANO, M., and MAGOUN, H. W. (1953) A neural basis of the anaesthetic state. *Arch. Neurol. Psychiat.* **69,** 519.

FULTON, J. F. (1951) *Decompression Sickness*, W. B. Saunders, Philadelphia and London.

GELFAN, S. and TARLOV, I. M. (1955) Differential vulnerability of spinal cord structures to anoxia. *J. Neurophysiol.* **18,** 170.

GELLHORN, E. and HAILMAN, H. (1943) The effects of anoxia on sense organs. *Fed. Proc.* **2,** 122.

GELLHORN, E. and HAILMAN, H. (1944) The parallelism in changes of sensory function and electroencephalogram in anoxia and the effect of hypercapnia under these conditions. *Psychosom. Med.* **6,** 23.

GESELL, R. (1923) The chemical regulation of respiration. *Physiol. Rev.* **5,** 551.

GOODMAN, L. and GILMAN, A. (1943) *The Pharmacological Basis of Therapeutics,* Macmillan, New York.

GREEN, J. B. (1861) *Diving With and Without Armour.* (Ref. UNSWORTH 1960.)

HALDANE, J. S. and PRIESTLY, J. G. (1935) *Respiration*, Clarendon Press, Oxford.

HALL, A. L. and KELLY, H. B. (1962) Exposure of human subjects to 100 per cent oxygen at simulated 34,000 feet altitude for 5 days. Report No. NMC-TM-62-7. U.S. Naval Missile Center, Point Mugu, California.

HESSER, C. M. (1963) Measurement of inert gas narcosis in man. *Proceedings 2nd Symposium on Underwater Physiology*, National Academy of Sciences, Washington, Publication 1181, p. 202.

HILL, L. and GREENWOOD, M. (1906) The influence of increased barometric pressure on man. *Proc. Roy. Soc.* B **77,** 442.

HILL, L. and MCLEOD, J. J. (1903) The influence of compressed air on respiratory exchange. *J. Physiol.* **29,** 492.

HILL, L. and PHILLIPS, A. E. (1932) Deep sea diving. *J. Roy. Nav. Med. Serv. London* **18,** 157.

HILL, L., DAVIS, R. H., SELBY, R. P., PRIDHAM, A. and MALONE, A. E. (1933) Deep diving and ordinary diving. Report of committee appointed by the Admiralty.

HYDEN, H. (1962) The neuron and its glia—a biochemical and functional unit. *Endeavour* **21,** 144.

JULLIEN, G., ROGER, A. and CHATRIAN, G. E. (1953) Preliminary report on variations of the E.E.G. of the cat at various air pressures. *Riv. Neurol.* **23,** 357.

KANE, H. F. (1940) The use of helium and oxygen in the treatment of asphyxia neonatorum. *Am. J. Obstet. Gynecol.* **40,** 140.

KIESSLING, R. J. and MAAG, C. H. (1962) Performance impairment as a function of nitrogen narcosis. *J. Appl. Psychol.* **46,** 91.

LAMBERTSEN, C. J. (1955) Respiratory and circulatory actions of high oxygen pressure. *Proceedings of the 1st Underwater Physiology Symposium*, Publication 377, National Academy of Sciences, Washington, p. 25.

LAWRENCE, J. H., LOOMIS, W. F., TOBIAS, C. A. and TURPIN, F. H. (1946) Preliminary observations on the narcotic effect of xenon with a review of values for solubilities of gases in water and oils. *J. Physiol.* **105**, 197.

LAZAREV, N. V. (1941) The biological action of gases under pressure. *Leningrad Naval Medical Acad.* and in *Farmakolg. i toksikolog.* **6**, 29 (1943).

LAZAREV, N. V., LYUBLINA, Y. I. and MADORSKAYA, R. Y. (1948) Narcotic action of xenon. *Fiziol. Zhur. S.S.S.R.* **34** (1), 131.

LEON, H. A. and COOK, S. F. (1960) A mechanism by which helium increases metabolism in small mammals. *Amer. J. Physiol.* **199**, 243.

LEVY, L. and FEATHERSTONE, R. M. (1954) The effect of xenon and nitrous oxide on *in vitro* guinea pig brain respiration and oxidative phosphorylation. *J. Pharmacol.* **110**, 221.

LILLIE, R. S. (1916) Physico-chemical Theory of Anesthesia. *The American Year-Book of Anesthesia and Analgesia* (Edited by F. H. MCMECHAN) Surgery Publ. Co.

LINAWEAVER, P. (1961) Use of helium–oxygen mixtures in mixed gas SCUBA oxygen limits. "Operation pulse beat". Report on Project NS 186-201, Experimental Diving Unit, Washington.

LINDSLEY, D. B. (1958) *Reticular Formation of the Brain*, Churchill, London.

LLOYD, D. P. C. and MCINTYRE, A. K. (1949) On the origins of dorsal root potentials. *J. gen. Physiol.* **32**, 409.

MANSFIELD, G. (1909) Narkose und Sauerstoffmangel. *Pflüg. Arch. ges. Physiol.* **129**, 69.

MARSHALL, J. M. (1950) Nitrogen narcosis in frogs and mice. U.S. Naval Research Report No. CO-503-2. July 1950–Jan. 1951.

MARSHALL, J. M. (1951) Nitrogen narcosis in frogs and mice. *Amer. J. Physiol.* **166**, 699.

MARSHALL, J. M. and FENN, W. O. (1950) The narcotic effects of nitrogen and argon on the central nervous system of frogs. *Amer. J. Physiol.* **163**, 733.

MERWARTH, C. R. and SIEKER, H. O. (1961) Acid–base changes in blood and cerebrospinal fluid during altered ventilation. *J. Appl. Physiol.* **16**, 1016.

MEYER, H. H. (1899) Theorie der Alkoholnarkose. I. Mitt welche Eigenschaft der Anasthetika bedingt ihre narkotische Wirkung. *Arch. F. exper. Path. u. Pharm.*, **42**, 109.

MEYER, J. S., GOTOH, F. and TAZAKI, Y. (1961) Continuous recording of arterial pO_2, pCO_2, pH and O_2 saturation *in vivo. J. Appl. Physiol.* **16**, 896.

MEYER, K. H. and GOTTLIEB-BILLROTH, H. (1920) Theory of narcosis by inhalation anaesthesia. *Zeits. f. physiol. Chem.*, **112**, 55.

MEYER, K. H. and HOPFF, H. (1923) Narcosis by inert gases under pressure. *Zeits. f. physiol. Chem.* **126**, 288.

MICHAELIS, M. and QUASTEL, J. H. (1941) Site of action of narcotics in respiratory processes. *Biochem. J.* **35**, 518.

MILES, S. (1962) *Underwater Medicine*, Staples Press.

MILES, S. and MACKAY, D. E. (1959) The nitrogen narcosis hazard and the self contained diver. *Medical Research Council Report.* R.N. Personnel Research Committee.

MILLER, S. L. (1961) A theory of gaseous anesthetics. *Proc. Nat. Acad. Sci.* **47**, 1515.

MORRIS, L. E., KNOTT, J. R. and PITTINGER, C. B. (1955) Electroencephalographic and blood gas observations in human surgical patients during xenon anaesthesia. *Anesthesiology* **16**, 312.

MOXON, W. (1881) The Croonian Lectures on the influence of the circulation upon the nervous system. *Brit. Med. J.* **1**, 491–497, 583–585.

REFERENCES

OLSEN, N. S. and KLEIN, J. R. (1947) Effect of cyanide on the concentration of lactate and phosphates in brain. *J. Biol. Chem.* **167**, 739.
OVERTON, E. (1901) *Studien über die Narkose*, G. Fisher, Jena.
PAULING, L. (1961) A molecular theory of general anesthesia. *Science* **134**, 15.
PITTINGER, C. B. (1962) Mechanisms of anesthesia. I. Xenon as an anesthetic. *Proceedings of the 22nd International Congress of Physiological Sciences, Leiden, Holland*, Exerpta Medica Foundation, London, vol. 1, ch. 2, p. 531.
PITTINGER, C. B. and KEASLING, H. H. (1959) Theories of narcosis. *Anesthesiology* **20**, 204.
PITTINGER, C. B., CONN, H. L., FEATHERSTONE, R. M., STICKLEY, E., LEVY, L. and CULLEN, S. C. (1956) Observations on the kinetics of transfer of xenon and chloroform between blood and brain in the dog. *Anesthesiology* **17**, 523.
PITTINGER, C. B., FAULCONER, A., KNOTT, J. R., PENDER, J. W., MORRIS, L. E. and BICKFORD, R. G. (1955) Electro-encephalographic and other observations in monkeys during xenon anesthesia at elevated pressures. *Anesthesiology* **16**, 551.
PITTINGER, C. B., FEATHERSTONE, R. M., CULLEN, S. C. and GROSS, E. G. (1951) Comparative *in vitro* study of guinea pig brain oxidations as influenced by xenon and nitrous oxide. *J. Lab. Clin. Med.* **38**, 384.
PITTINGER, C. B., FEATHERSTONE, R. M., GROSS, E. G., STICKLEY, E. E. and LEVY, L. (1954) Xenon concentration changes in brain and other body tissues of dog during inhalation of the gas. *J. Pharmacol. Exper. Therap.* **110**, 458.
PITTINGER, C. B., MOYERS, J., CULLEN, S. C., FEATHERSTONE, R. M. and GROSS, E. G. (1953) Clinopathologic studies associated with xenon anesthesia. *Anesthesiology* **14**, 10.
POULTON, E. C., CARPENTER, A., and CATTON, M. J. (1963) Mild nitrogen narcosis? *Brit. Med. J. Dec.* **2**, 1450.
QUASTEL, J. H. and WHEATLEY, A. H. M. (1932) Narcosis and oxidations of the brain. *Proc. Roy. Soc.* B **112**, 60.
RASHBASS, C. (1955) The unimportance of carbon dioxide in nitrogen narcosis. *Medical Research Council Report*. R.N. Personnel Research Committee.
READ, J. (1958) Effect of a combined treatment with 8-ethoxycaffeine and argon on the roots of *Vicia Faba*. *Nature* **181**, 616.
RICE, J. A. (1960) Polarographic studies of the cat's auditory cortex. *J. Neurophysiol.* **23**, 350.
RIKKL, A. V. and KRIVOSHEENKO, N. K. (1948) The effect of increased atmospheric pressure on the higher nervous function of the dog. In *Trudy Nauchoi Sessii, Posviashchennoi Tridtsatiletiiu Velikoi Oktiabr'skoi Sotsialisticheskoi Revolutsii* (Edited by A. V. TRIUMFOV), Leningrad, p. 62.
RINFRET, A. P. and DOEBBLER, G. F. (1961) Physiological and biochemical effects and applications. *Argon, Helium and the Rare Gases*, G. A. Cook, Interscience, vol. 2, ch. 19.
ROGER, A., CABARROU, P. and GASTAUT, H. H. (1955) E.E.G. changes in humans due to changes in surrounding atmospheric pressure. *Electroenceph. clin. Neurophysiol.* **7**, 152.
ROSENBERG, L. C. and RAMSDELL, R. C. (1957) Changes in intellectual function associated with nitrogen narcosis. Thesis, Key West, Fla.
RUSSEK, M. (1962) Histotoxic hypoxia. *Proceedings of the 22nd International Congress of Physiological Sciences, Leiden, Holland, International Congress Series* **47**, Exerpta Medica Foundation, London.
SCHREINER, H. R. (1962) Biological effects of the rare gases. *Proceedings of the 22nd International Congress of Physiological Sciences, Leiden, Holland, International Congress Series* **47**, Exerpta Medica Foundation, London.

REFERENCES

SEARS, D. F. (1962) Mechanisms of anesthesia. III. Role of lipid molecules in anesthesia and narcosis. *Proceedings of the 22nd International Congress of Physiological Sciences, Leiden, Holland,* vol. 1, ch. 2, p. 540. Exerpta Medica Foundation, London.

SEARS, D. F. and FENN, W. O. (1957) Narcosis and emulsion reversal by inert gases. *J. Gen. Physiol.* **40,** 515.

SEARS, D. F. and GITTLESON, S. M. (1961) Narcosis of paramecia with xenon. *Fed. Proc.* **20**(1), 142.

SEUSING, J. (1961) The problem of depth intoxication. *Wehrmed. Mitt.* No. 10, 150.

SEUSING, J. and DRUBE, H. (1960) The importance of hypercapnia in depth intoxication. *Klin. Wschr.* **38,** 1088.

SEVERINGHAUS, J. W. (1959) *A Symposium on pH and Blood Gas Measurement* (Edited by R. F. WOOLMER), Churchill, London, p. 126.

SHILLING, C. W. and WILLGRUBE, W. W. (1937) Quantitative study of mental and neuromuscular reactions as influenced by increased air pressure. *U.S. Nav. Med. Bull.* **35,** 373.

SOUTH, F. E. and COOK, S. F. (1953) Effect of helium on the respiration and glycolysis of mouse liver slices. *J. Gen. Physiol.* **36,** 513.

THESLEFF, S. (1956) The effect of anesthetic agents on skeletal muscle membrane. *Acta physiol. scand.* **37,** 335.

THOMAS, J. J., NEPTUNE, E. M. and SUDDUTH, H. C. (1963) Toxic effects of oxygen at high pressure on the metabolism of D-glucose by dispersions of rat brain. *Biochem. J.* **88,** 31.

UNSWORTH, I. P. (1960) Nitrogen narcosis. Some aspects. *St. Mary's Hospital Gazette, London* **66,** 272.

UNSWORTH, I. P. (1963) Personal Communication.

VAN HARREVELD, A. (1944) Survival of reflex contraction and inhibition during cord asphyxiation. *Amer. J. Physiol.* **141,** 97.

VAN HARREVELD, A. (1946) Asphyxial depolarisation of the spinal cord. *Amer. J. Physiol.* **147,** 669.

VAN HARREVELD, A. and MARMONT, G. (1939) The course of recovery of the spinal cord from asphyxia. *J. Neurophysiol.* **2,** 101.

VERWORN, M. (1903) *Die Biogenhypothese,* Jena.

VERWORN, M. (1912) *Narcosis. Harvey Lectures,* J. B. Lippincott Co., Philadelphia.

WOOD, W. B. (1962) Personal Communication. Dept. of Medicine, University of North Carolina, U.S.A.

WOOD, W. B., LEVE, L. H., and WORKMAN, R. D. (1962) Ventilatory dynamics under hyperbaric states. Report 1–62. U.S.N. Experimental Diving Unit, Washington D.C.

WORKMAN, R. D., BOND, G. F. and MAZZONE, F. H. (1962) Prolonged exposure of animals to pressurised normal and synthetic atmospheres. Report No. 374. U.S.N. Medical Research Lab., Conn.

WULF, R. J. and FEATHERSTONE, R. M. (1957) A correlation of Van der Waals constants with anaesthetic potency. *Anesthesiology* **18,** 97.

WYKE, B. D. (1960) *Principles of General Neurophysiology Relating to Anaesthesia and Surgery,* Butterworths, London.

YAMAGUCHI, T. (1961) Electrophysiological studies on the mechanism of effects of anaesthetics on the isolated frog muscle fibre. *J. Fac. Sci. Hokkoido Univ. (Ser. IV, Zool.)* **14,** 522.

ZETTERSTROM, A. (1948) Deep sea diving with synthetic gas mixtures. *Milit. Surg.* **103,** 104.

ZETTERSTROM, A. (1949) *J. Indust. Hyg.* **31,** abstract section 23–24.

AUTHOR INDEX

Adolfson, J.   3, 11, 13, 69
Albano, G.   8–11, 34, 35, 69, 70, 80
Arduini, A.   79
Arduini, M. G.   79

Baddely, A.   95
Barnard, E. E. P.   10, 11, 35, 71
Bean, J. W.   xvi, 21, 31, 32, 34, 39, 69, 90
Behnke, A. R.   2, 15, 16, 21, 22, 27, 86
Bennett, P. B.   6, 17, 35, 36, 38, 40, 45, 48, 49, 51, 66, 68, 70–80, 83, 87–91, 92, 94, 95–99
Bert, P.   2
Bickford, R. G.   67–69
Bishop, P. O.   46
Bjurstedt, H.   86
Bond, G. F.   15, 37, 38
Brazier, M. A. B.   58
Brink, F.   24, 25, 27
Brooks, C.   46
Buhlman, A. A.   34–36, 51, 54
Burker, K.   58
Burnett, W. A.   27
Buros, O. K.   7
Butler, T. C.   xiv, 42

Cabarrou, P.   10, 33, 34, 36, 39, 66–67, 80
Carpenter, A.   14, 97
Carpenter, F. G.   14, 17, 43–45, 46, 58, 60, 80, 83
Case, E. M.   4, 6, 10, 11, 12, 22, 27, 41, 86
Catton, M. J.   14, 97
Chang, H.   79
Chatrian, G. E.   66, 67, 80
Chun, C.   45, 82
Ciulla, C.   8–11, 34, 35, 69, 80
Clutton-Brock, J.   72

Collins, W. F.   79
Conn, H. L.   29
Cook, S. F.   50, 60, 61
Courtin, R. F.   67
Cousteau, Y.   3
Criscuoli, P. M.   8–11, 34, 69, 80
Cross, A. V. C.   75
Cullen, S. C.   15–17, 29, 36
Culpin, M.   20

Damant, G. C. C.   2, 21
Danielli, J. F.   42
Davis, H. S.   79
Davis, R.   2
Dean, R. B.   50
Dillon, W. F.   79
Doebbler, G. F.   15
Dossett, A. N.   35, 70, 71, 92
Drube, H.   34
Dunn, J. A.   13
Dusser de Barenne, J.   38, 52

Ebert, M.   47
Eccles, J. C.   46
End, E.   15, 86

Faulconer, A.   67–69
Featherstone, R. M.   xiv, 16, 25, 27, 29, 58, 59, 70
Fenn, W. O.   18, 44, 63, 97
Ferguson, J.   25, 27
Ferris, E. B.   29
Frada, G.   69
Frank, G. B.   82
Frankel, J.   61
Frankenhaeuser, M.   11, 12
French, J. D.   79
Fulton, J. F.   xvi

AUTHOR INDEX

Gastaut, H. H.   66, 67, 80
Gelfan, S.   46
Gellhorn, E.   76
Gesell, R.   41
Gilman, A.   xiv
Gittleson, S. M.   18
Glass, A.   6, 36, 66, 68, 72–74, 80
Goodman, L.   xiv
Gotoh, F.   38, 55, 84
Gottlieb-Billroth, H.   16
Graff-Lonnevig, V.   11, 12
Green, J. B.   1
Greenwood, M.   31
Gross, E. G.   15–17, 29, 36, 70

Hailman, H.   76
Haldane, J. B. S.   4, 6, 10–12, 22, 27, 32, 41, 86
Hall, A. L.   13
Hempleman, H. V. H.   10, 11, 35
Hesser, C. M.   11, 12, 39
Hill, L.   2, 20, 31
Hopff, H.   16, 80
Hornsey, S.   47
Howard, A.   47
Hyden, H.   84

Ikels, K. G.   22, 94, 95

Jullien, G.   66, 67, 80

Kane, H. F.   50
Keasling, H. H.   42, 63
Keller, H.   34, 87
Kelly, H. B.   13
Kidd, D. J.   35, 70, 71
Kiessling, R. J.   7, 8, 12, 27, 74, 93
Klein, J. R.   61
Knott, J. R.   16, 36, 37, 66–68, 72
Krivosheenko, N. K.   65

Lambertsen, C. J.   40
Lawrence, J. H.   15, 17, 94

Lazarev, N.   17, 43
Leon, H. A.   61
Leve, L.   35, 36
Levy, L.   29, 58, 59, 70
Lillie, R. S.   82
Linaweaver, P.   40
Lindsley, D. B.   76
Lloyd, D. P. C.   46
Loomis, W. F.   15, 17, 94
Lyublina, Y. I.   17

Maag, C. H.   7, 8, 12, 27, 74, 93
Mackay, E.   8, 14
Madorskaya, R. Y.   17
Magoun, H. W.   79
Mansfield, G.   58
Marmont, G.   90
Marshall, C. S.   38, 52
Marshall, J. M.   18, 26, 41, 44, 64, 65, 70, 80, 87, 93
Mazzone, F. H.   37, 38
McCulloch, W. S.   38, 52
McIntyre, A. K.   46
Mcleod, J. G.   46
Mcleod, J. J.   2
Merwarth, C. R.   38
Meyer, H. H.   21, 22, 27, 58, 65, 83, 84
Meyer, J. S.   38, 52, 84
Meyer, K. H.   16, 17, 80
Michaelis, M.   58
Miles, S.   8, 13, 14
Miller, S. L.   27, 64
Molle, W. E.   29
Morris, L. E.   16, 36, 37, 66–68, 72
Motley, E. P.   2, 16, 21
Moxon, W.   19
Moyers, J.   16
Muelbaecher, C.   xiv, 27

Neptune, E. M.   61, 62
Nims, L. F.   38, 52

Olsen, N. S.   61
Overton, E.   21, 22, 27, 58, 65, 83, 84

AUTHOR INDEX

Pauling, L. 27, 63
Pender, J. W. 68, 69
Phillips, A. E. 2, 20
Pittinger, C. B. 16, 29, 36, **37, 42,** 63, 66, 67, 70, 72
Posternak, J. M. 24, 25, 27
Poulton, E. C. 14, 97, 99
Priestly, J. G. 32

Quastel, J. H. 58

Ramsdell, R. C. 6
Randt, C. T. 79
Rashbass, C. 6–7, 32–33
Ray, P. 92
Read, J. 47
Rice, J. A. 80
Rikkl, A. V. 65
Rinfret, A. P. 15
Roger, A. 66, 67, 80
Rosenberg, L. C. 6
Russek, M. 50
Ryder, H. W. 29

Sanders, H. D. 82
Schneiderman, H. A. 61
Schreiner, H. 48, 50, 93
Sears, D. F. 18, 27, 63, 82, **83**
Seusing, J. 34, 36
Severin, G. 86
Severinghaus, J. W. 38
Shilling, C. W. 3, 4, 5, **12,** 93
Sieker, H. O. 38
South, F. E. 50, 61

Stickley, E. E. 29, 70
Sudduth, H. C. 61, 62

Tarlov, I. M. 46
Tazaki, Y. 38, 52, 84
Thesleff, S. 82
Thomas, J. J. 61, 62
Thomson, R. M. 2, 16, 21
Tobias, C. A. 15, 17, 94
Trotter, C. 10, 11, 35
Turpin, F. H. 15, 17, 94

Unsworth, I. 27, 84, 85

Van Harreveld, A. 46, 90
Verworn, M. 58
Verzeano, M. 79
Visscher, M. B. 50

Wheatley, A. H. M. 58
Willgrube, W. W. 3, 4, 5, 12, 93
Wood, W. B. 29, 35–36, 70, 71, 93
Workman, R. D. 35–38
Wulf, R. J. 25, 27
Wyke, B. D. 82

Yamaguchi, T. 82
Yarbrough, O. D. 15, 22, 27, 86
Young, D. R. 61

Zetterstrom, A. 22, 86

SUBJECT INDEX

Absolute pressure, definition of xiv
Absorption, rate of 29, 74, 92, 93
Acclimatization 3–6, 14, 27, 74–75, 96, 98
Action potentials 42–44; Fig. 8
  and sodium conductivity 82
Activity, thermodynamic 24–27, 43
Adsorption
  and narcotic potency 27
  site of 47, 83
Alcohol 14
Alcoholic intoxication 1, 3
Alpha activity 66
  blocking of 72–75
Altitude xvi, 13
Alveolar carbon dioxide 31–34
Alveolar ventilation 35
Anaesthesia
  and lipid solubility 21
  and metabolic mechanisms 58–63
  and Van der Waals constants 25–26
  definition of xiii–xiv
  due to xenon 16–17, 36–37, 66–69
  permeability theory in 82
Apprehension 3, 14
Argon 4, 15, 17, 20
  and cortical available oxygen 49–57
  and flicker fusion frequency 76
  and frog brain waves 26, 65
  and frog reflex 26, 44
  and frog sciatic nerve 44
  and spinal synapses 45–47, Figs. 9–10
  and Van der Waals Constants 26
  effect of CO_2 on 41
  effect on brain CO_2 38–40

effect on evoked potentials 76–79
effect on irradiated bean roots 47
effect on *Neurospora crassa* 48
oil/water phase reversal in 63
oxygen consumption in 60–61
partition coefficient of 22–23
protection by Frenquel 88
Arithmetic tests
  and alpha blocking 72–75
  and Frenquel 89
  at different N_2/O_2 mixtures 8–10
  effect of work on 13
  errors in 3–13, 72–75
  helium and 4, 95, 96
  neon and 93, 94
  rapid compression and 92, 93
Arterial
  carbon dioxide 37
  oxygen 37
Aspirin 90–91
Asphyxia
  and spinal synapses 46
  as mechanism of narcosis 58
  at synapse 46, 83
Assisted ventilation 34, 51–54
Atmospheric pressure, definition of xiv

Bean roots, radio sensitivity of 47
Bends xvi, 95
Blood
  and inert gas exchange 29, 68
  carbon dioxide 37–38
  oxygen 37–38, 69
  pH 31, 37–38
  pooling of 19
  pressure 20, 68

Brain
- and carbon dioxide tension 33, 38–40
- blood pooling in 19
- electrical activity and lactic acid 61
- metabolism in, 58–61
- oxygen 48–57
- saturation by inert gases 29, 70
- synapses in 44–45, 60, 78, 79, 81
- waves in cats 52–57, 66, 76–80
- waves in frog 26, 65
- waves in man 36, 66–67, 69, 72–75
- waves in monkey 68
- waves in rats 70

Breath-holding as narcosis test 20–21
Breathing resistance, increase in 13, 31–36, 50

Caisson workers 2, 97
- decompression sickness in xvi

Cancellation test
- letters 6–7
- numbers 4–5, 7
- with Frenquel 89

Carbachol 83, 90–91
Carbon dioxide 2, 4
- alveolar 31–34
- and psychometric tests 12, 32, 41
- and rate of compression 69–72
- blood tensions 37, 38
- in brain 38–40, 71–72, 96
- narcosis 84
- production of 61
- retention of 13, 21, 31–41, 69, 97
- synergistic effect of 12, 41, 53–54, 69, 70, 84, 97

Cell membrane
- and sodium conductivity 82
- permeability of 27, 82

Clathrate formation 27, 63–64
Claustrophobia 2
Compressed air narcosis, signs and symptoms 1–15, 92, 93, 97
Compression
- effect on brain CO_2 31–34, 38–40, 96
- rate of, 13–14, 32, 69–72, 92, 96

Concentration of inert gas in brain 29, 70
Conceptual reasoning test 7–8
Consciousness, loss of, in air 2–3, 71
Convulsions
- electroshock 44
- oxygen xvi, 90, 91

Cortical
- carbon dioxide 38–40, 52–57
- evoked potentials 39, 49, 51–57, 66, 76–80
- oxygen 48–57

Critical narcotic threshold 65, 66, 72–80
Cyanide 50, 79
Cyanosis 47
Cyclopropane xiii, 60, 67, 79

Decompression sickness xvi, 95
Demarcation potential 43
Density, effect of 13, 34–36, 39, 50
Depolarization 44
Depth, metric equivalents xvi
Digit substitution test 6
Dizziness
- and rapid compression 4, 69, 92
- due to helium 40, 96
- due to krypton 15

Doriden 90–91

Electrocardiogram 20
Electroconvulsive shock 43–44
Electroencephalogram 36, 66–67, 69
- and alpha blocking 72–74
- carbon dioxide and 49–57
- evoked potentials and 49–57, 66, 76–80
- in cats 52–57, 66, 76–80
- in frogs 26, 65
- in monkey 68
- in white rats 70
- xenon and 66–69

Electromyograms 45
Electroshock as index of narcosi 35, 70–71, 87–91

SUBJECT INDEX 113

Emotional stability 3
Emulsion reversal 63-64
Equivalents, metric xvi
Ether xiii, 67, 79, 82
Euphoria 3, 6, 10, 92
Evoked potentials 39, 49-57, 66, 76-80
Excitatory synaptic transmitters 82-83
Experience in compressed air 3-5, 14, 93
Exposure, length of 6, 7, 27, 72-75

Fatigue 3, 14
Flicker fusion frequency 75-76, 89
Frenquel 87-91
Frog
   brain waves in 26, 65
   narcosis in 18
   reflex preparation in 26, 41, 44, 93
   sciatic nerve in 44

Gauge pressure, definition of xiv
Glial cells 84
Goldstein-Scleerer test 6

Haemoglobin 68
Handwriting, deterioration of 3, 10
Helium 20, 29, 87
   and alpha blocking 72
   and brain carbon dioxide 38-40
   and cortical available oxygen 49-51
   and frog reflex 41, 44, 93
   and hydrate formation 64
   and sciatic nerve 44
   and spinal synapses 45-46
   ease of ventilation in 34, 36
   effect of CO_2 on 41
   effect on evoked potentials 76-79
   effect on frog electroencephalogram 65
   effect on oxygen convulsions 39, 40
   effect on performance 4, 95-98
   increased carbon dioxide output in 61
   increased oxygen consumption in 61
   in deep diving 86-87, 95-98
   in irradiated bean roots 47
   in *Neurospora crassa* 48
   low narcotic potency of 15-18, 26, 40, 41, 43, 95-98
   oil/water phase reversal in 63
   oxygen consumption in 60-61
   partition coefficient 22-23
   production of carbon dioxide 61
Histotoxic hypoxia 46, 47-58, 64, 76, 83-85, 90
Hydrates 63-64
Hydrogen
   and hydrate formation 64
   effect on irradiated bean roots 47
   effect on performance 4, 22, 86
   in deep diving 86
   partition coefficient of 22-23
Hypercapnia 38-40, 70, 96, 97
Hyperexcitability
   at different N_2/O_2 p.p. 69
   of neurones 66, 69, 80
Hyperventilation 51-53
   and psychometric tests 33
Hypocapnia 39-40, 51-53
Hypoventilation 34-36
Hypoxemia 69, 79
Hypoxia xvi, 3, 76
   at synapse 46, 47, 83
   histotoxic 46, 47-57, 61, 76, 79, 90
   on evoked potentials 79

Inert gases
   definition of xiv, 42
   oil solubility of 22-23
   saturation by 27, 29-30, 70
   threshold 76
Inert gas narcosis
   cause of 19, 41
   signs and symptoms 15-18
Inhibition of frog reflex 26, 41, 44, 93
Inhibitory synaptic transmitters 82-83
Intelligence 4, 6, 10

SUBJECT INDEX

Ionic movement   44, 81–82
Isonarcotic potency   25

Krypton   15, 17, 26, 28, 43
  dizziness in   20
  effect of density   36
  effect on irradiated bean roots   47
  effect on *Neurospora crassa*   48

Lactic acid   61–63
Laughter   3
Learning   1, 7–8, 14, 92, 93
Leptazol   90–91
Lipid solubility
  of anaesthetics   21
  of inert gases   26, 94, 95
  of nerve tissue   22
Loquacity   3
Lung
  damage   xvi
  ventilation   31–37
  volume   xv

Manual dexterity   4, 6–8, 89, 94–99
Maximum breathing capacity   36
Mechanical dexterity   7–8, 14, 89, 94–99
Mechanisms of the narcosis   42–64
Megimide   90–91
Memory   7–8, 14, 69
Metabolism   50, 58–63, 84
Methedrine   90–91
Metric equivalents   xvi
Meyer–Overton theory   21–22, 27, 58, 65–66, 83, 84
Mirror drawing test   11–12
Mitochondria   83
  and oxidative phosphorylation   58–59
Molar concentration   17, 25, 43, 66, 80
Molecular volume   26–27
Molecular weight   17, 27–28, 36, 48
Motivation   1, 14

Narcosis
  definition of   xiii–xvi, 35
  rate of onset   29–30, 70, 92–93

Narcotic concentration
  in non-synaptic pathways   25
  in synaptic pathways   25
Narcotic potency   21–29
Neon
  and hydrate formation   64
  and Van der Waals constants   26
  effect on *Neurospora crassa*   48, 93
  narcotic potency of   15, 26, 93–95
  partition coefficient of   22–23
Nerve
  block of conduction in   44–47, 60
  lipid solubility in   22
  oxygen consumption of   60
Neurospora crassa   48, 93
Nitrogen   2, 4, 10, 16, 20–21, 34
  and alpha blocking   72–75
  and arterial saturation   29
  and blood pH, pCO_2, pO_2 37–38
  and carbon dioxide retention   33
  and changes in spinal transmission   45
  and cortical available oxygen   49–51
  and frog brain waves   26, 65
  and frog reflex   26, 41, 44, 93
  and frog sciatic nerve   44
  and spinal synapses   45–46
  blanket theory   13
  critical molar concentration   17, 25, 43, 66, 80
  E.E.G. in   26, 36–37, 65–80
  effect of CO_2 on   41
  effect of different N_2 p.p.   8, 11, 14, 34–36, 69, 70
  effect of different O_2 p.p.   8, 11–14, 34–36, 62, 69, 70, 97
  effect of time at pressure   6, 7, 29–30, 74–75, 93
  effect on brain CO_2 39–40
  effect on evoked potentials   76–80
  effect on irradiated bean roots   47
  effect on *Neurospora crassa*   48
  hyperexcitability due to   66, 69, 80
  lactic acid production in   61–63
  partition coefficient of   22–23

SUBJECT INDEX

Nitrogen narcosis
  blood CO_2 in   37, 38
  blood O_2 in   37, 38
  blood pH in   37, 38
  effect of O_2 variation on   8, 11, 34-35, 62, 69, 70, 97
  in insects   16-17, 97
  oil/water phase reversal in   63
  oxidative phosphorylation in   61, 62
  protection by Frenquel   88-91
  rate of onset of   6, 11, 17, 27, 29-30, 70, 92-93
Nitrous oxide   xiii, 66, 68-69, 72, 79

Oil solubility
  and asphyxial mechanisms   58
  and narcotic activity   44
  and narcotic potency   26-27
  and Van der Waals Constants   27
  of anaesthetics   21
  of inert gases   22-23, 26, 43, 94, 95
Onset of narcosis   6, 11, 17, 27, 29, 30, 70, 72, 92, 93
  critical pressure of   7, 14, 97
Oxygen
  and brain carbon dioxide   12, 39, 97
  and hypoventilation   34
  as cause of narcosis   21
  blood tensions   37, 38, 69
  consumption of   60-61
  convulsions   xvi, 21, 40, 90-91
  cortical available   49-57
  effect of decreased p.p. of   6
  effect on evoked potentials   66
  effect on nitrogen narcosis   8-10, 34-35, 48, 95, 97
  hyperexcitable effect of   66, 69, 80
  lack of   xvi, 6, 48, 84
  pharmacological studies in   90-91
  poisoning   xvi, 21, 40, 90, 91
  synergistic action of   6, 12, 41, 51, 97
  tension   49-58, 69
  utilization in ether   58
  with irradiation and narcosis   47

Paraesthesia due to helium   40
Paramecia, effect of xenon on   18
Partial pressures, definition of   xv
Partition coefficients
  and bean root irradiation   47
  of inert gases   22-23
Peripheral nerve, transmission in   42-46, 60
Permeability of ions   27, 82-84, 89
pH   31, 37-38, 84
Pharmacological studies   87-91
Phase reversal, oil/water   63-64
Phenacetin   90-91
Physostigmine   90-91
Polarographic oxygen   48-57
Porteus maze test   6
Pressure
  absolute, definition of   xiv
  atmospheric, definition of   xiv
  definition of   xiv
  gauge, definition of   xiv
  metric equivalents   xvi
  partial, definition of   xv
Psychological stress as cause of narcosis   2, 20-21, 27
Psychometric tests   3-15, 92-99
  and alveolar CO_2 32-33
  and Frenquel   89
  rate of deterioration of   29, 72, 92, 93
Pulse rate   20, 68
Pump sodium   84, 85
Pyruvate   58-60

Rate
  of compression   4, 13, 32, 69-71, 92-93
  of narcosis onset   6, 11, 17, 27, 29, 30, 70, 72, 92-93
Reaction time test   4-5, 7-8, 11-12, 76, 92, 93, 96-98
Recovery from narcosis   6, 11, 14, 17, 96
Reflex, impairment of   26, 41, 45, 65, 93
Respiratory embarrassment   35-36, 50

SUBJECT INDEX

Retention of carbon dioxide 31–41, 69, 96
Reticular formation 72, 76–80, 85

Saturation by inert gases 29–30, 70, 93
Scopolamine 90–91
Sodium
  extrusion pump 84–85
  ions 44, 81–82, 85
Solubility coefficients 22–23, 94–95
Spinal synapses 45–47
Sulphur hexafluoride 48
Synapse 81–85
  and activity function 25
  and brain 45, 60, 77–80
  conduction failure in 46–47, 63, 90
  inhibitory 45, 90
  metabolism in 60
  spinal 45–47
  transmission 42, 64, 81
Synaptic vesicles 83
Synergistic
  action of CO_2 12, 41, 53–54, 69, 84, 97
  action of oxygen 6, 12, 35, 41, 51, 97

Temperature changes of 27
Thermal conductivity of helium 61
Thermodynamic activity 24–27, 43
Tissue
  carbon dioxide 31, 38–40
  oxygen 49–57

Valsava manoeuvre 27, 92

Van der Waals Constants 25–27
Venous
  carbon dioxide 37, 38
  oxygen 37, 38
Ventilation
  assistance of 34, 51–53
  controlled rate of 53–57
  decreased 34–36
Vital capacity 20
Voice distortion
  due to helium 87
  due to krypton 15

Water solubility in anaesthesia 21
Work
  effect on arithmetic tests 13–14
  effect on narcosis 3, 13–14, 96
Writing, deterioration in 3, 10

Xenon
  anaesthesia 16, 20, 68–69, 72
  and blood gases 37
  and guinea pig brain metabolism 58
  and oxidative phosphorylation 59
  E.E.G. in 66–67
  effect of density of 36
  effect on irradiated bean roots 47
  effect on *Neurospora crassa* 48
  narcotic properties of 15–18
  on blockade of axons 60
  on pyruvate consumption 60
  oxygen consumption in 60–61
  saturation in brain of 29